P9-EEK-542

RIVERSIDE COMMUNITY COLLEGE
1916

HOME MECHANIX
GUIDE TO SECURITY

HOME MECHANIX GUIDE TO SECURITY

PROTECTING YOUR HOME, CAR, & FAMILY

BILL PHILLIPS

John Wiley & Sons, Inc.
New York • Chichester • Brisbane • Toronto • Singapore

This text is printed on acid-free paper.

Published by John Wiley & Sons, Inc.

Library of Congress Cataloging in Publication Data:

Phillips, Bill
Home mechanix guide to security : protecting your home, car, & family / Bill Phillips.
p. cm.
Includes index.
ISBN 0-471-58893-8 (pbk.)
1. Burglary protection. 2. Security systems. 3. Fire prevention. I. Mechanix illustrated. II. Title.
TH9705.P49 1994
643'.16—dc20 93-10303

Printed in the United States of America

10 9 8 7 6 5 4 3 2 1

To my nephew Antonio

List of Trademarks Used in *Home Mechanix Guide to Security*

Abloy Disklock is a registered trademark of Abloy Security Locks.

Bodyguard is a registered trademark of Medeco Security Locks.

Carbrella is a registered trademark of Carbrella Motoring Accessories.

D-10 Series is a registered trademark of Medeco Security Locks.

First Alert is a registered trademark of Pittway Corporation.

Fortress is a registered trademark of VSI Donner.

Gardall is a registered trademark of Gardall Safe Corporation.

The Guardian is a registered trademark of Twenty First Century International Fire Equipment and Services Corp.

Heath is a registered trademark of Heath Zenith.

Hidden Assets is a registered trademark of Ellen Pekarsky Enterprises, Inc.

Intermatic is a registered trademark of Intermatic Incorporated.

Kwikset is a registered trademark of Black and Decker.

Lexan is a registered trademark of G.E. Plastics Group.

LoJack is a registered trademark of LoJack Corporation.

Lucite is a registered trademark of Auburn Plastic Engineering.

Plexiglas is a registered trademark of Rohm & Haas Company.

Preso-matic is a trademark of Preso-Matic Lock Company.

Primus is a registered trademark of Schlage Lock Company.

Quick-Vent is a registered trademark of M.A.G. Engineering and Manufacturing, Inc.

Redi-Exit is a registered trademark of Karsulyn Corporation.

Residential Protection Specialist is a registered trademark of International Association of Home Safety and Security Professionals, Inc.

Steelguard is a registered trademark of Medeco Security Locks.

UL is a registered trademark of Underwriters Laboratories Inc.

X-10 is a registered trademark of X-10 (USA) Inc.

X-10 Powerhouse is a registered trademark of X-10 (USA) Inc.

Foreword

Readers of *Home Mechanix* magazine know that it offers a unique approach to maintaining, improving, and enhancing the value of your home. We call it "managing your home in the '90s." This approach differentiates *Home Mechanix* from other how-to home improvement magazines. It reflects our belief that facing the challenges of home ownership requires more than knowing how to fix what breaks and build or install amenities. To manage your home effectively, it's important to see a bigger picture and develop proficiency with tools in addition to those you find in a workshop.

Home Mechanix' managing-your-home tools are informational: expert advice, how-to instruction, product information, and news about home-related trends. With these tools, all homeowners can gain the decision-making power to make the most of their resources of time, money, and skill. Exercising power to set and accomplish goals, after all, is what managing is all about. Where is it more appropriate or important to do it wisely than in matters connected to your home?

Security is an increasingly important aspect of managing your home in the '90s. That's not because crime is more prevalent or shocking than ever before, as the tabloids and TV would lead us to believe. Rather, today's homeowners have greater awareness of the need to address security concerns and invest in more effective options for protecting themselves, their families, and their property from potential threats. When we at *Home Mechanix* decided to address these concerns through the column "Home Security," we were privileged to enlist Bill Phillips to guide our efforts.

Bill is an experienced security professional and a widely published writer in his fields of expertise. While his credentials and hands-on experience are impressive, it's his point of view that makes *Home Mechanix Guide to Security: Protecting Your Home, Car, & Family* an especially useful book.

The title itself helps to define *security* in a way that is most relevant to your concerns. It extends the concept of home security beyond narrow strategies for keeping intruders out of your home, and addresses matters of automotive security and personal safety. We at *Home Mechanix* view family vehicles as an extension of the home, and we believe that those who take an active interest in managing their homes are equally interested in managing their cars. Many of the issues, including security, are similar for houses and cars. Bill's effort to link the concept of security with personal safety places the emphasis where it belongs: on protecting yourself and the members of your family from harm—including potential threats from would-be intruders, home fires, accidents, and amateurish security measures that you might adopt yourself without being aware of the inherent hazards.

Bill's innovation is to treat anti-break-in, anti-theft measures for houses and cars; potential fire emergencies; security lighting; self-defense; and insurance protection as integral interrelated matters central to the issue of home security. This book is the only one we know of that treats homeowners' security concerns from this perspective.

The text provides both general and specific information on the latest equipment, including alarm systems, locks, safes, fire sprinklers, lighting devices, and video and intercom systems designed to enhance security. Through the clear instructions and detailed illustrations in this book, you can learn to install and use this equipment properly to discourage attacks on your property and family, and to warn you when danger is imminent. There is also a great deal of valuable consumer information on how to assess your security needs and options, where to find security equipment at the best prices, and how to find security installers to do work that you choose not to do yourself. Along with Bill Phillips' broad expertise, *Home Mechanix Guide to Security* is further enriched by information gained from interviews with other experts from various security specialties.

For me, the most satisfying aspect of this book has been the opportunity to collaborate with Bill in presenting information on the subject of security in greater depth and more detailed focus than is possible in the briefer magazine format. With its broad scope grounded in the fundamentals, the book will stand as a useful guide to homeowners' security concerns, even as the technologies that Bill has written about here continue to evolve. Bill's efforts, both in the book and in the pages of *Home Mechanix* magazine, reflect the best of what the managing-your-home-in-the-'90s approach has to offer.

MICHAEL CHOTINER
Editor-in-Chief
Home Mechanix

Acknowledgments

While writing this book, I received help from many people. Although it isn't possible for me to mention everyone by name, I'm thankful to all of them. I'd like to give special thanks to *Home Mechanix* Editor-in-Chief Michael Chotiner and John Wiley & Sons Editor Judith McCarthy for their encouragement, criticism, and helpful advice.

I'd also like to gratefully acknowledge the assistance of Michael Morris, Director of Group Publishing Projects for *Home Mechanix* and *Popular Science,* Jane Jordan Browne of Multimedia Product Development, Assistant Managing Editor Angela Murphy of John Wiley & Sons, Nancy Marcus Land and Maryan Malone of Publications Development Company, Senior Editor Steve Ross (formerly of John Wiley & Sons), and the following safety and security experts: Postal Inspector John Brugger of Washington, DC; Harold Gillotti, President of Hi-Tech Industries Inc., Fort Wayne, IN; Martial Artist Jeffrey J. Kelly of Ukiah, CA; Chief Fire Inspector Gregory Martin of Erie, PA; Alexander Murray, President of Malverne Alarms, Malverne, NY; Gerald A. O'Rourke, CPP, President of Strategic Controls, Inc., New York, NY; Senior Lighting Designer Julia Rezek of Grenald Associates, Ltd., Culver City, CA.

Special thanks also to the following associations and companies who generously provided illustrations and technical information:

Abloy Security Inc.

Aiphone Corporation

Alarm Accessories Ltd.

BRK Electronics

Canadian Centre for Justice Statistics

CHB Industries

The David Levy Company, Inc.

Dicon Systems

Directed Electronics

Falcon Eye, Inc.

Gardall Safe Corporation

Generation Two

Group Three Technologies

Heath Zenith Reflex Brand Group

Hidden Assets

InteLock Corporation

Intermatic Incorporated

International Association of Home Safety and Security Professionals, Inc.

Karsulyn Corporation

Kryptonite Corporation

Leslie-Locke, Inc.

Lillian Vernon Corporation

LoJack Corporation

M.A.G. Engineering & Manufacturing

Majestic Company

Marvin Windows & Doors

Medeco Security Locks

Molvan Enterprises Inc.

MSI Mace

NAPCO Security Systems, Inc.

National Fire Protection Association

National Fire Sprinkler Association, Inc.

Norden Lock Company, Inc.

Preso-Matic

Progress Lighting

Questech International, Inc.

Quorum International, Ltd.

Rudolph-Desco Company, Inc.

Sentry Group

Topper Hardware Inc.

Transcience

Twenty First Century International Fire Equipment & Services Corporation

United States Department of Justice

United States Fire Administration

United States Postal Service

Unity Systems Inc.

Vantage Technologies Inc.

Velux-America Inc.

VSI Donner

X-10 (USA) Inc.

Most importantly, I'd like to thank Oscar Carr, Ruby Carr, Gloria Glenn, Daniel Phillips, and Michael Phillips for their help and inspiration, without which this book would not have been written.

BILL PHILLIPS

Erie, Pennsylvania
August 1993

Contents

Foreword by Michael Chotiner, Editor-in-Chief, *Home Mechanix* vii

Acknowledgments ix

Introduction 1

Chapter 1 Thinking about Protection 3

The Threat of Crime 4
The Threat of Burglary 4
The Threat of Fire 5

Chapter 2 Strengthening Doors and Windows 7

How Burglars Can Open Your Doors 8
Choosing a New Door 15
Securing Your Windows 17
Recommended Resources 24

Chapter 3 Selecting a Good Lock 25

Grades of Door Locks 26
Types of Door Locks 26
Lock Cylinders 45
Minimizing Your Keys 46
Recommended Resources 52

Chapter 4 The Safest Safes 53

Types of Protection 54
Degrees of Protection 54
Where and How to Install Safes 55
Special Features to Look For 57
Specialty Safes 57
Getting a Good Buy 57
Alternatives to a Safe 59
Recommended Resources 61

Chapter 5 Electronic Security for Today and the Future 63

Burglar Alarms 64
Professional versus Do-It-Yourself Alarms 68
Home Automation Systems 72
Recommended Resources 78

Chapter 6 Lighting for Security, Safety, and Beauty 83

Light Sources 85
Light Controllers 85
Preventing Accidents 86
Recommended Resources 91

Chapter 7 Closed Circuit Television Systems 95

How CCTV Works 96
Video Intercoms 98
Recommended Resources 105

Chapter 8 Preventing and Surviving Home Fires 109

Causes and Cures 110
Smoke Detectors 111
Fire Extinguishers 111
Escape Ladders 113
Fire Sprinkler Systems 113
Surviving a Home Fire 118
What to Do after a Fire 119
Recommended Resources 121

Chapter 9 Keeping Thieves Out of Your Car 123

Anti-Theft Devices 124
Car Alarms 125
Stolen-Vehicle Retrieval Systems 129
Combating Carjackers 130
Recommended Resources 133

Chapter 10 Personal Safety at Home and Away 135

Physical Self-Defense Training 136
Rape Prevention 136
Self-Defense while Traveling 137
Learning from the L.A. Riots 138
Recommended Resources 141

Chapter 11 Insurance Considerations 143

Understanding Your Policy 144
How Much Building Protection Is Enough? 144
Personal or Home-Business Property 145
Special Disasters 145
Available Discounts 145
Federal Crime Insurance Program 146
Recommended Resources 149

Chapter 12 Choosing and Using Security Professionals 151

Locksmiths 153
Home-Alarm Systems Installers 154
Physical Self-Defense Instructors 155
Security Consultants 156

Chapter 13 The Safe and Secure Home 159

Surveying a House 160
Home Safety and Security Checklist 161
Surveying an Apartment 161
Final Thoughts 164

Appendix A Ways to Use X-10 Products 165

Appendix B *HM* Sourcelist 197

Glossary 209

Index 213

About the Author 217

HOME MECHANIX
GUIDE TO SECURITY

Introduction

You've probably heard these three myths:

1. If burglars want to get in badly enough, you can't keep them out.
2. The only way to keep burglars out is to use expensive high-tech devices.
3. Without a lot of training or a weapon, you can't thwart a determined attacker who's larger and stronger than you are.

Often such myths are perpetuated by books and articles written by well-meaning journalists who use crime victims or "former burglars" as primary sources of information. But neither a burglar nor a crime victim is a security expert.

Contrary to movies and television crime dramas, the criminals who break into homes are not James Bond clones. Most home burglars know little about defeating locks or electronic security systems. Many rely on just one or two quick, quiet, and simple entry techniques, and they search for homes at which they can use those methods.

Home security articles and books that are based on advice from burglars often contain misleading, incomplete, and potentially dangerous information. In many cases, the advice one burglar gives to thwart his favorite break-in technique can make it easier for other burglars to use their techniques.

Good security advice isn't based on scattered facts or personal anecdotes, but rather on a comprehensive understanding of security measures, self-defense techniques, security hardware, and electronic systems. I've never been burglarized, attacked, or had a car stolen, nor have I ever been a criminal, but I have had years of training and experience in locksmithing, safe and vault work, alarm-system installation, martial arts, and security consulting. I've also lived in some of the highest crime areas in the United States, including New York City and Detroit. I know it *is* possible for people to stay safe and protect their property without spending a lot of money.

I had been writing primarily for law enforcement officers and other security professionals when Michael Chotiner, who was then Executive Editor of *Home Mechanix,* asked me to help with the magazine's new "Home Security" column. I jumped at the offer because I saw it as an opportunity to clear up some myths about security, and to come up with innovative ways for homeowners to be safer and save money on security products and services.

The column was well received, but I wanted to give more how-to information than the limited magazine space allowed. That's why I decided to write this book. Although this book came about as a result of the "Home Security" column, most of the information here hasn't appeared in the column. To make this book as comprehensive, authoritative, and practical as possible, I spent over a year writing it while consulting with some of the most knowledgeable safety and security experts.

As you read this book, you'll learn to think like a security expert. You'll learn how to assess intelligently your safety and security needs, and how best to meet those needs. If you or someone you know has been victimized by household crime or fire, you'll probably discover within these pages why the incident occurred and how it might have been prevented.

The subjects covered include: how burglars defeat locks, combating car thieves, psychological and physical self-defense, why some safes aren't safe, deciding whether you need a burglar alarm, installing alarms, home automation systems, making an apartment safer, lighting for security, getting more homeowner's insurance for less money, making doors and windows stronger, choosing and using security professionals, and using closed circuit television systems.

Each chapter includes do-it-yourself projects, money-saving tips, and practical advice that you can use right now, whether you live in a large city or small town. To help you find the best security products at the lowest prices, in Appendix B you'll find the addresses and toll-free phone numbers of most major product manufacturers and distributors. Nearly every sentence of this book is written in nontechnical language, so you should have no trouble following the advice and instructions. But I've included a glossary that should be helpful if you find an unfamiliar word.

Chapter 13 includes a comprehensive home safety and security checklist to guide you in surveying your home. You may want to glance at it now, but you'll benefit more by using the checklist after you've read the whole book.

Because all the topics in this book are interrelated, you'll get the most out of it by reading it from beginning to end. But even if you skip around, you'll discover new ways to protect your home, your car, and your family.

1

Thinking about Protection

As I pointed out in the Introduction, there are many myths about staying safe. People who believe them either spend a great deal of money and time trying to create an impenetrable fortress, or make no attempt to insure their safety. By understanding your real risks, you'll know why you can be safe from crime and fire without turning your home into a prison and without emptying your bank account. That's what this chapter is all about.

THE THREAT OF CRIME

The first step to safeguarding yourself, your family, and your property is to realize no one is immune from crime, and that you need to take smart proactive security measures. These facts from the United States Department of Justice may be helpful:

- In 1992, 1 in 4 households were burglarized or had a member who was the victim of rape, robbery, assault, or theft.
- Households with an annual income of between $15,000 to $25,000 are more likely to be burglarized or experience a violent crime than are households earning over $50,000 per year.
- Nearly 22 percent of burglaries occur through unlocked doors or windows.

As the numbers show, neither the very poor nor the very rich are the only ones who need to be concerned about crime. But I don't want to spend a lot of time discussing crime statistics; they don't accurately account for individual actions. Most of the statistics group the small number of people who take smart security measures with the much larger number of people who do little to prevent crime. With a good security strategy, you can greatly reduce your risk of being victimized.

THE THREAT OF BURGLARY

You may have heard someone say, "I don't worry about burglars because I don't have anything worth stealing." But burglars don't break into homes because valuables are in it. They break into a place because they *think* valuables might be in it. What's junk to one burglar may be valuable to another—depending on how successful the burglar is.

But often burglars do much more than just take things. Sometimes they destroy property and maim or kill people. Even if nothing is damaged and no one gets hurt, a break-in can leave you feeling violated and scared. Some burglary victims never again feel safe at home.

Much of this book focuses on how to keep burglars out of your home. Because burglars are the most adept at getting into homes, the things you do to keep them out will also keep many other types of criminals out.

Types of Burglars

There are three basic types of burglars: professional, semi-professional, and amateur. Each type works differently. As the name implies, professionals make a living as burglars. Like other professional people, when they wake up in the morning, they have a cup of coffee and read the newspaper. Instead of reading the stock reports or sports pages, however, professional burglars look for leads on places to burglarize. They read the obituaries to learn when a wealthy family might be away at a funeral. They look for stories about lottery winners, or people leaving town on business or vacation trips.

Professionals have a network of business contacts. They work with parking lot attendants, real estate agents, and delivery persons to get leads. They also work with "fences," seemingly reputable businesses that buy stolen goods (often pawn shops and other stores that sell used items).

Professional burglars are the type most often depicted in movies and television crime dramas. Whoopi Goldberg's character in the movie *Burglar* is a good example. Like that character, professionals carefully choose targets, and then use all types of ruses, disguises, and special tools to gain entry. Professional burglars may spend weeks, or even months, planning a particular break-in. After gaining entry, they quickly go through select rooms to find specific items, such as jewelry, collectibles, or cash. Such deliberation and expertise makes professionals hard to stop.

But unless you have a home safe full of valuables, and you give media interviews about it, no professional burglar is likely to think about you or your home. To professionals, time is money. They have no interest in randomly choosing homes to break into, because they know most homes contain few valuables. Professional burglars primarily target businesses and very expensive homes.

Semi-professional and amateur burglars are the types most of us need to be concerned about. Both are opportunists, and neither has extensive knowledge of locking devices or security systems. Semi-professionals know a few simple entry techniques (see Chapters 2 and 3 for examples), and actively look for homes in which to use those techniques. Amateurs break into a place only when confronted with an overwhelmingly simple opportunity—such as an open door when no one is home.

THE THREAT OF FIRE

A good home security strategy must include fire safety considerations. As you'll discover throughout this book, many things people do while trying to protect themselves from crime inadvertently increase their risk of being victimized by fire. For example, sometimes they install door locks that make it hard for occupants to escape when necessary.

In addition to high crime rates, the United States and Canada have the highest fire-death rates in the industrialized world. Each year fire kills more Americans and Canadians than all major natural disasters combined, including floods, hurricanes, tornadoes, and earthquakes. All of the security suggestions in this book take fire safety into account, and comprehensive advice on avoiding and escaping fires is given in Chapter 8.

2

Strengthening Doors and Windows

Whenever I'm asked about home security, the question seems to be concerned mainly with door locks. Rarely does anyone think to ask about making the doors themselves stronger. I've been in homes where people were asking me to recommend a lock for a door that looked as though it would fall down if anyone knocked on it too hard.

It's important to understand that a lock is just a device that fastens a door to one side of a door frame. Using a good lock on a thin-paneled door or on a door with weak hinges is like using a heavy-duty padlock to secure a paper chain. Before worrying about a good lock, be sure you have a strong door and frame.

If burglars can't get in through your doors, where do you think they might try next? To keep burglars out, you need to secure *all* of your points of entry—windows, skylights, skyroofs, and other wall and roof openings, as well as the doors. In this chapter we'll discuss how burglars might get through your home's openings, how to make them more secure, and how to choose and install new doors and windows.

HOW BURGLARS CAN OPEN YOUR DOORS

Burglars are not invited guests. Don't allow them to enter your home as though they were. Be aware of burglars' secrets for getting past doors, and what you can do to keep them out.

Removing the Hinges

If a door's hinges can be seen from the exterior side, a burglar may be able to remove the hinges and open the door without touching the lock. Most door hinges consist of two metal leaves (or "plates")—each with "knuckles" on one edge—and a pin that fits vertically through the knuckles when they're aligned and holds the leaves together. The hinge pins often can be pulled out with little difficulty, and the door then becomes disconnected from the door frame. Burglars who remove a door in that way can place the door back on its hinges on the way out, and you may never know how they got in (and your insurance company may not want to pay your claim). One way to prevent this type of entry is to use hinges with nonremovable pins—pins that are either welded in place or secured by a set screw or retaining pin.

If you don't want to replace your door hinges, you can install hinge enforcers. These small metal devices attach to the hinge and the door frame to block the door's removal even when the hinge pins have been removed. A package of hinge enforcers costs less than $5.

Prying Off Stop Molding

If your door's hinges can't be seen from outside, your next concern should be your door stop molding. Stop moldings are the protruding strips (usually about 1/2 inch thick) that are installed on three sides of a door frame—the lock side, the hinge side, and the header (top). They stop the door from swinging too far when you're closing it. Depending on which way the door swings, a person standing outside the door will be able to see either the hinges or the stop molding.

Some stop moldings are simply thin wooden strips tacked to the frame and can be easily pried off. By removing the lock side strip, a burglar exposes the bolt and thus makes it easier to attack the lock. To solve the problem, you can remove the stop moldings and reinstall them using wood glue and nails so that they can't be pried off. When you buy or make a new door and frame, be sure that its stop molding is milled as an integral part of the jamb.

Kicking Doors Down

The most common way burglars get through doors is by kicking them. If either the strike plate on the door jamb or the lock edge of the door is weak, a strong kick will knock the door open. Short of getting a new door, the best way to solve a weak-door problem is to install door reinforcers (Figure 2–1). They usually cost less than $20 each.

One type of door reinforcer is a U-shaped metal unit designed to wrap around the door edge near the lock. Designs are available for doors with one or two locks. To install this type of reinforcer, first remove the locks from your door. Position the unit so that the lock holes are fully exposed, and screw it firmly into place. Then install the locks. (See Figures 2–2 and 2–3.)

Weak door frames can also be strengthened. A popular reinforcer for door frames is the high-security strike box, a heavy-gauge steel box with long screws or rods that protrude through the door jamb and into a wall stud. The strike box is stronger than the more commonly used thin, flat strike plates that are fastened only to the jamb, using small wood screws.

THE PROBLEM

DOORS AND FRAMES ARE WEAK

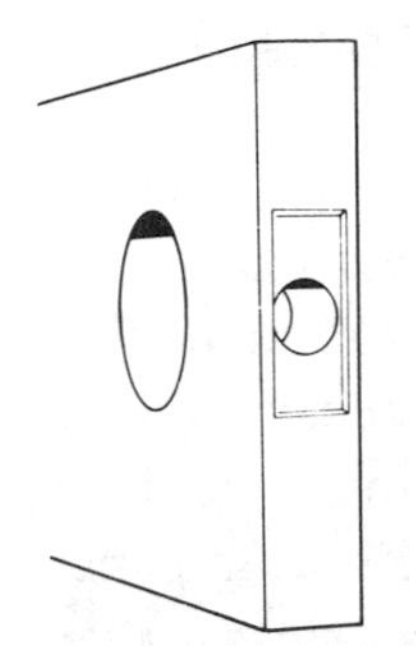

LOCK INSTALLATION HOLES WEAKEN DOOR IN THIS AREA CAUSING IT TO BREAK OUT WHEN DOOR IS KICKED OR FORCED OPEN.

TYPICAL DOOR PREPARATION FOR DEADBOLT OR LOCKSET NEEDS REINFORCEMENT

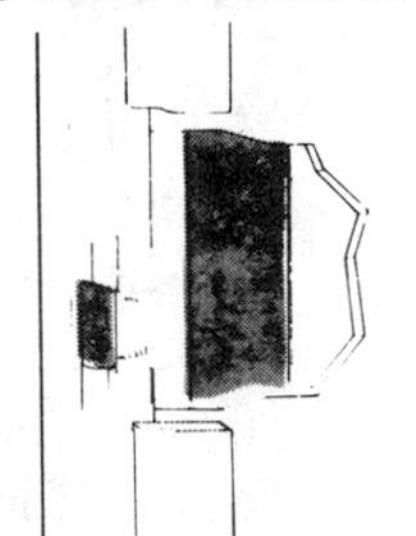

TYPICAL STRIKE IS ATTACHED TO SOFT WOOD JAM WITH ONLY 2 SHORT SCREWS.

DOOR FRAME IS EASILY BROKEN WHEN DOOR IS KICKED OR PRIED NEEDS HI-SECURITY STRIKE

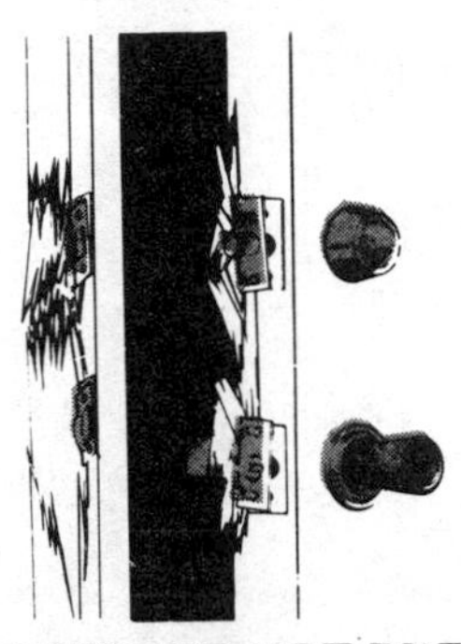

THE LOCKS DID NOT BREAK THE DOOR AND FRAME BROKE NEEDS REINFORCEMENT

LOCKS ARE ONLY AS STRONG AS THE WOOD DOOR AND FRAME

THE SOLUTION

DOOR REINFORCERS AND HI-SECURITY STRIKES

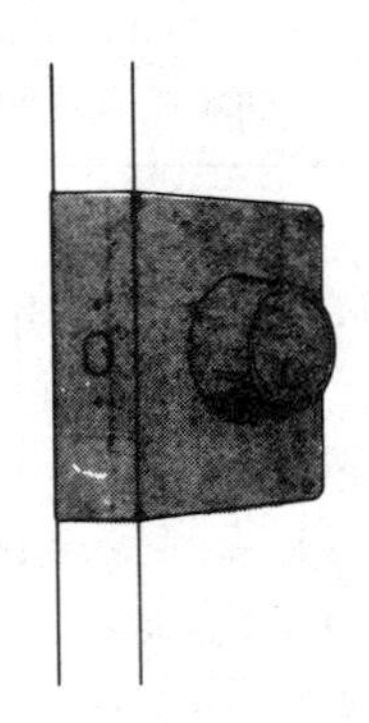

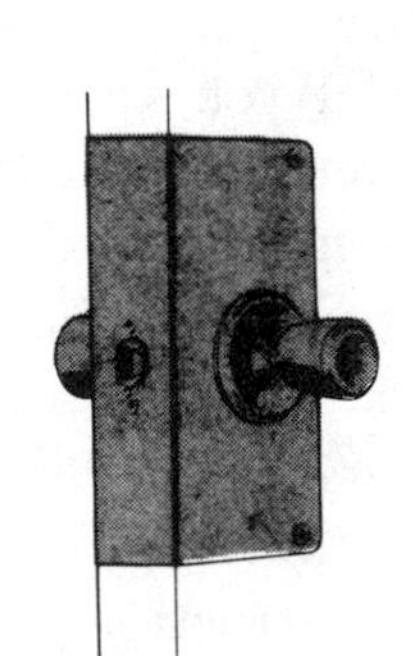

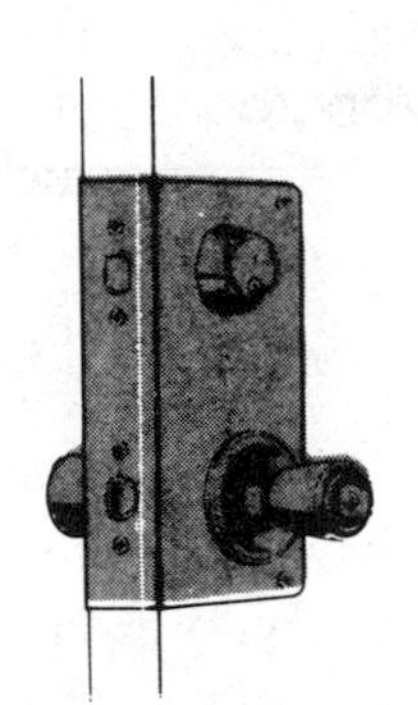

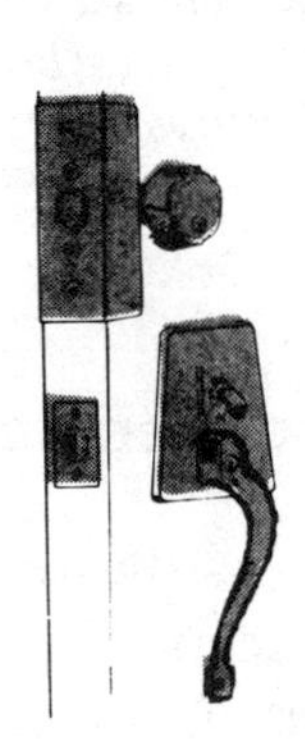

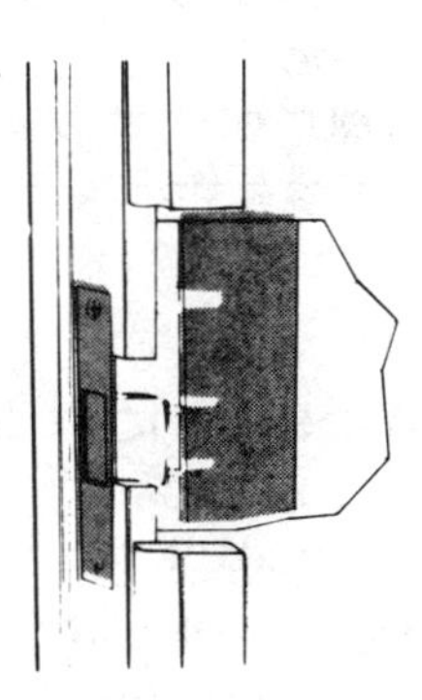

CHANNEL ENCASES LOCK, BOLT & DOOR IN ONE SOLID UNIT
CONVERT MORTISE LOCKS TO KEY-IN-KNOB LOCKS
PREVENT OR REPAIR DOOR DAMAGE & BEAUTIFY YOUR DOOR

HEAVY DUTY BOX STRIKE ANCHORS TO STUD WITH 4 HARDENED SCREWS
A MUST FOR ALL DEADBOLTS

MAKE THE DOOR AND FRAME AS STRONG AS THE LOCK

Figure 2–1. A wide variety of door reinforcers are available to strengthen weak or damaged doors. (Courtesy of M.A.G. Engineering & Manufacturing Co., Inc.)

Figure 2–2. Some door reinforcers are made to be used on doors with deadbolt locks. (Courtesy of M.A.G. Engineering & Manufacturing Co., Inc.)

Impersonating Other People

No matter how much hardware you have on your doors, it won't keep out burglars who trick you into letting them in. Burglars often pretend to be delivery persons, police officers, or meter readers, and trusting people immediately open the door for them. Always ask to see identification, or ask that the full address on the delivery be read to you before you open the door. If your door has no glass pane, install a wide-angle door viewer that lets you see anyone standing outside. (See Figure 2–4.) Better models allow you to see in several directions. (See Figures 2–5 and 2–6.) Door viewers range in price from under $5 to over $50.

Don't depend on a door chain to let you see who's at your door. Door chains provide a false sense of security. A person who intends to break in can easily snap the chain by pushing sharply against the door.

Prying Open Sliding Glass Doors

A sliding glass door (sometimes called a "patio" door) usually consists of two glass panels (or "sashes") that slide along tracks. Doors of this type are especially vulnerable because their frames and locks are weak. A sliding glass door can be forced open by prying the sliding panel away from the door frame. You can thwart that entry technique by inserting a length of wood into the metal track that the door slides on. The wood acts as an obstruction; the door cannot be opened from the outside. You can easily reopen it from the inside by removing the wood. Various styles of sliding door barriers are sold in hardware stores and home improvement centers. (See Figure 2–7.) Some models are designed to let you keep the door open a few inches for ventilation. (See Figure 2–8.)

Burglars can defeat sliding glass doors by using a pry bar to lift the sliding sash out of its lower track. You can install screws or antilift plates at the top of the door to resist that entry technique. (See Figure 2–9.) These devices create an obstruction between the door and frame. You can buy a package of antilift plates for less than $5.

To install an antilift plate, first close and lock the door. Position the plate so that it is on the center of the frame and butts against the upper track, then screw the antilift plate into place. (You may need to use an awl to puncture screw holes into the frame.) An antilift plate isn't adjustable for clearance all along the track. Antilift screws also resist attempts to lift a sliding door out of its lower track, but they are adjustable to allow for door clearance all along

INSTALL-A-LOCK®

INSTALLATION INSTRUCTIONS

LISTED 76S1 DOOR LOCK ACCESSORY

NOTE: First, check to be sure this unit matches the backset, door thickness and hole size for your installation.

FOR INDENTED UNIT

Fig. 1

1. Drill holes in door using instructions furnished with lockset purchased, or, remove old lockset from door.

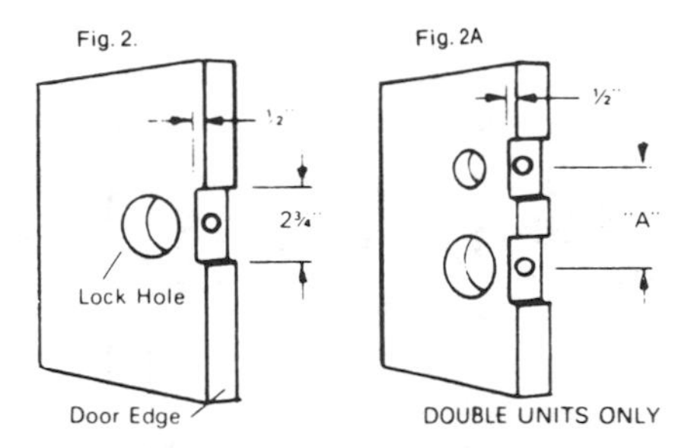

2. Cut out door edge as shown on center with lock hole. For double units (Fig. 2A), the "A" dimension (centers between lock holes) is 3⅝" or 4". See label on box.

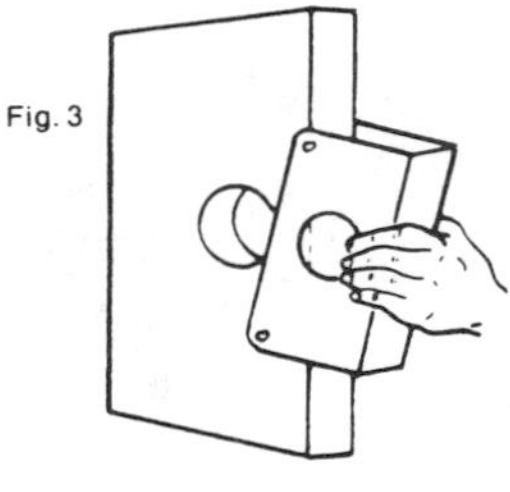

3. Slide unit over door.

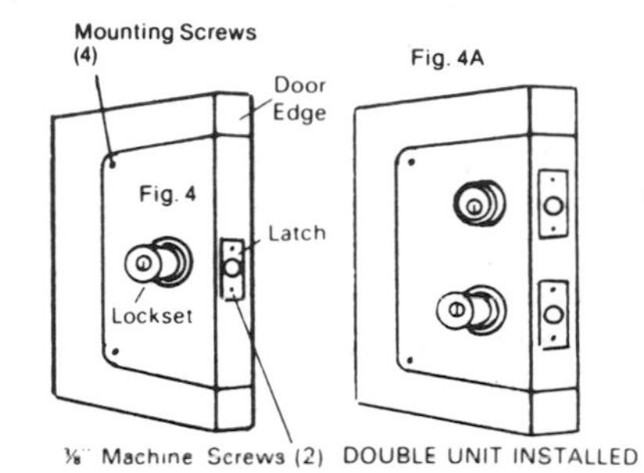

4. Install latch (with ⅜" machine screws provided), then lockset. Tighten lockset (not firm) to hold INSTALL-A-LOCK® in position. Push unit flush against edge of door, then tighten lockset. Install 4 mounting screws.

NOT FOR TIGHT FITTING DOORS: If INSTALL-A-LOCK® hits jamb when closing, remove and cut out 1/32" on door edge so INSTALL-A-LOCK fits flush with door edge.

5. For units supplied with strike: After unit is installed place strike on jamb centered with latch. Mark all around and cut out approximately 1/16" deep. Install with two 1" long screws furnished.

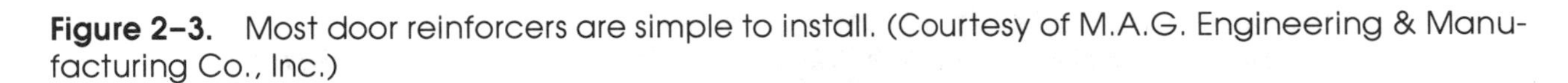

Figure 2–3. Most door reinforcers are simple to install. (Courtesy of M.A.G. Engineering & Manufacturing Co., Inc.)

Figure 2–4. A typical door viewer consists of two pieces designed to be screwed together through an opening in the door. (Courtesy of VSI Donner)

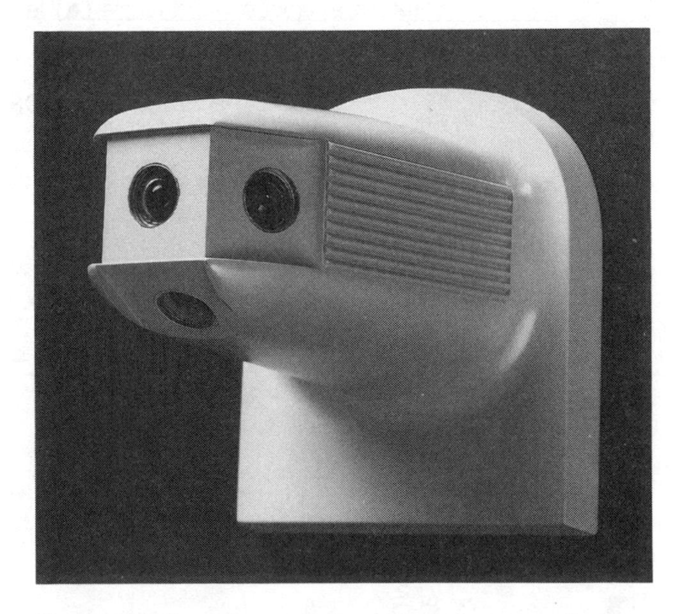

Figure 2–5. The most useful types of door viewers allow you to see in several directions. (Courtesy of Rudolph-Desco Co., Inc.)

DOOR SPY® DS-5

INSTALLATION INSTRUCTIONS

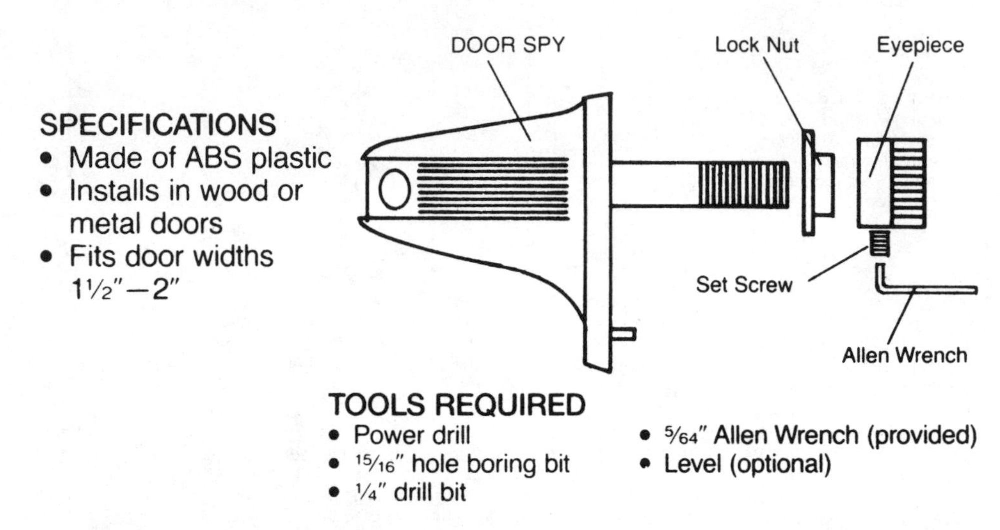

1. Detach TEMPLATE and place squarely in center of door at eye-level (a carpenter's level could be helpful). Tape in place.
2. Mark the center of both holes. Remove TEMPLATE.
3. Drill completely through door for top hole using 15/16″ hole boring bit.
4. Drill ¼″ deep for bottom hole using ¼″ drill bit.
5. Remove eyepiece (using Allen Wrench) and lock nut. Install DOOR SPY and tighten lock nut securely.
6. Return eyepiece with triangle (▲) facing upward making sure the view is straight ahead. Tighten with Allen Wrench.

A Product of DOOR SPY, INCORPORATED
11 East 47th Street, New York, NY 10017, (212) 754-0030

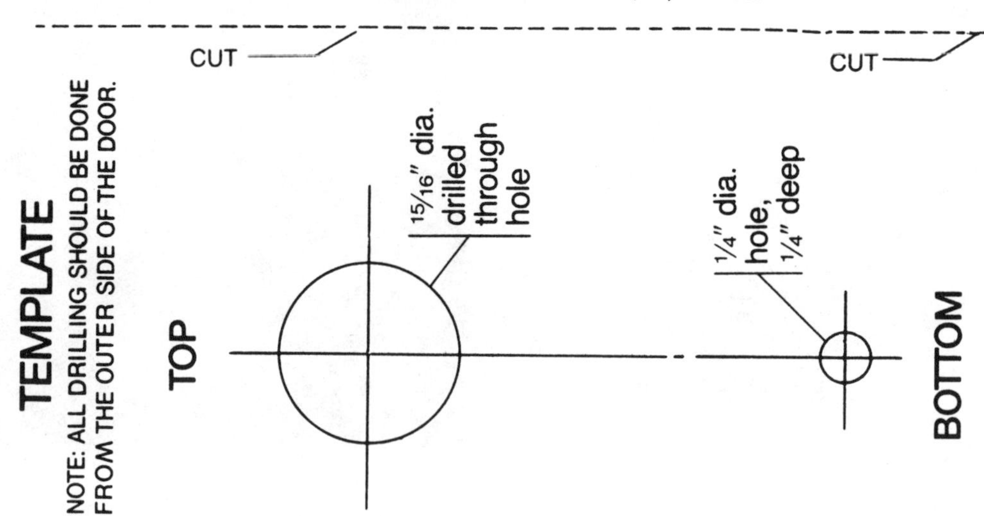

Figure 2–6. Only two holes need to be drilled in a door to install the Door Spy door viewer. (Courtesy of Rudolph-Desco Co., Inc.)

Figure 2–7. By placing a strip of wood along the length of a sliding door's track, you can prevent someone from being able to open the door from outside. (Courtesy of MSI Mace)

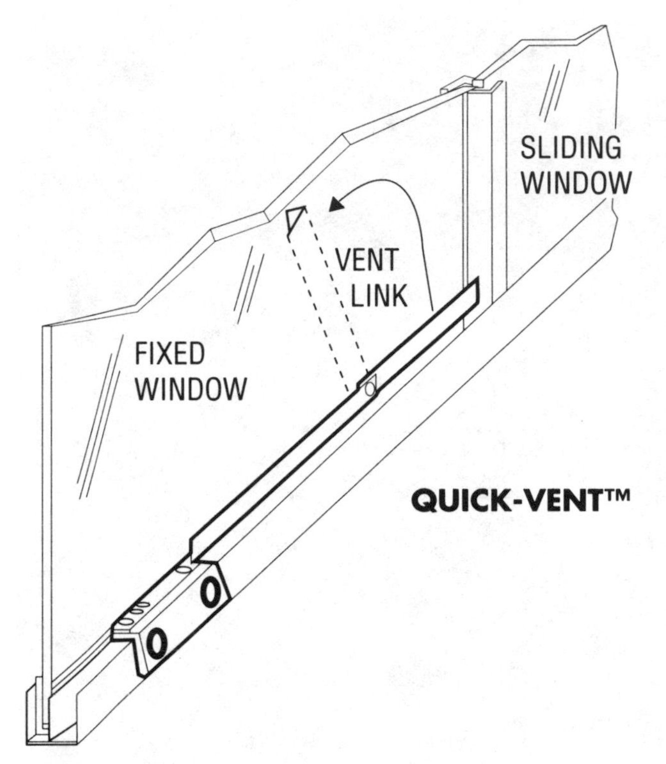

Figure 2–8. Some barriers for sliding glass doors allow you to keep the door open for ventilation. (Courtesy of M.A.G. Engineering & Manufacturing Co., Inc.)

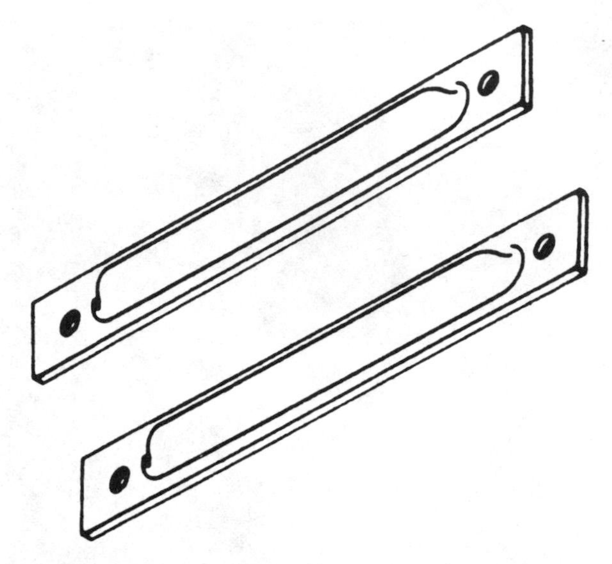

Figure 2–9. By fastening antilift plates to the top of a sliding glass door, you'll make it hard for a burglar to lift the door out of its frame. (Courtesy of VSI Donner)

the track. The screws cost about the same as the plates and need to be installed about every 12 inches along the length of the center of the upper track.

Breaking In through Garage Doors

Don't ignore your garage doors. Burglars know that a typical garage contains cars, bikes, lawn mowers, tools, and other easy-to-sell items. A garage that's attached to a home usually provides easy access to the home. The most secure main garage doors are made of steel, require an automatic door opener, and have no glass panels.

Panels of any material weaken a door, but glass panels are an especially poor feature in garage doors. They can be broken easily, and

How to Reinforce Garage Door Panels with Angle Iron

1. Using 3/4-inch by 3/4-inch angle iron that nearly spans the width of your garage door, position the angle iron so that it crosses the horizontal center of a row of panels.
2. Mark locations for screw holes along the bar about every 2 inches, if possible. Remove the bar from the door, and drill screw holes through the bar at the marked points.
3. Place the bar back into position on the door and use an awl to punch starter holes into the door. Then screw the angle iron into place.

they allow a burglar to see what's in the garage. If you have glass in a garage door, you might want to paint over the glass from the inside.

To reinforce panels of any material on a wood door, you can install angle iron. Even if a burglar breaks a panel, the angle iron will block entry.

In addition to securing the main door of a garage, you should reinforce any door that allows passage from the garage to your home. That "inside" door should be as secure as any exterior door. Burglars who are able to drive into your garage and enter your home through the garage entrance will be unseen while they load your possessions into their car.

On another topic, check your local building code. It may require that a door connecting a home to a garage be fire-resistant.

CHOOSING A NEW DOOR

In most homes, the style of door depends on the structure's architecture. You'll probably want doors that complement your home's design, but, regardless of the style, you'll need to decide on the materials of the door unit.

Typically, an exterior door is connected to its frame by metal hinges on one side and a lock bolt on the other. The frame consists of various sections: a head jamb (along the top), two side jambs, stop moldings on the top and sides, and a sill or threshold (along the bottom). Although the door and frame don't have to be made of the same material, they usually are. Commonly used materials include steel, wood, aluminum, fiberglass, polyvinyl chloride (PVC) plastic, and glass.

Steel doors offer the best protection against fire and break-in attempts. They also offer superior insulation, which helps keep energy costs down. Many steel doors are beautifully designed to look like expensive wood doors. Kalamen doors consist of metal wrapped around a wood core; they provide good security when installed with a strong frame.

Fiberglass, a strong material that can be made to look like natural wood, offers good resistance to warping and weathering; it's especially useful near pools and saunas or in damp areas. Some types of fiberglass can be stained and finished. Like fiberglass, PVC plastic is strong and isn't affected much by water. However, the plastic surface can be hard to paint.

Aluminum and glass are used together, mostly for sliding glass doors. Although glass can make a door look nice and admits light to the interior, it also makes the door less secure.

Among wood doors, the solid-core hardwood types are best. They consist of hardwood blocks laminated together and covered with veneer. A hollow-core door provides minimal protection; it consists of two thin panels over cardboardlike honeycomb material. You can recognize a hollow core door by knocking on it; it sounds hollow. If a burglar kicked a hollow-core door, his foot would go through it.

There's an easy way to reinforce a hollow-core door if aesthetics aren't important. You can clad the exterior side with 12-gauge (or thicker) sheet metal attached with 5/16-inch-diameter carriage bolts. The bolts should be placed along the entire perimeter of the door about 1 inch in from the door's four edges. Space the bolts about 6 inches apart and secure them with nuts on the interior side of the door. If after installing the metal you find that the door is too heavy to open and close properly, you may need to remove the hinges and install larger ones.

Another important factor affecting door strength is whether it's flush or paneled. A flush door is flat on both sides and is plain-looking. A paneled door has surfaces of varying thicknesses, and can be very attractive. The panels may be metal, wood, glass, or a combination of materials. Because the panels are usually thinner and weaker than the rest of

How to Replace an Exterior Door

Before prehung doors were introduced, the parts of the frame had to be cut, assembled, glued, and nailed together, and then the door had to be hung on the frame (using hinges). A prehung door unit comes already assembled on a frame and is ready to position and fasten in place. You can replace an exterior door with a prehung exterior door by doing the following:

1. Pry off the interior trim from around the four sides of the door. You'll have to work a pry bar along the entire length of each strip of trim. (Be careful not to scratch the wall around the door.)
2. Using a punch and mallet, remove the pins from each of the hinges on the door. Then remove the door (you may need a helper).
3. Use a saw to cut the threshold (the bottom of the door frame) into three sections. Then use a pry bar to remove each section of the threshold from the sill.
4. Pry the side jambs from the studs and the head jamb from the header. Take out any shims or nails that may be sticking out of the sill, studs, or header.
5. If necessary, nail plywood strips to the studs and header to make the dimensions of the rough opening about 1/2 inch larger than the overall size of the new door frame.
6. Run two parallel beads of caulk along the length of the sill. Then position the new door and door frame on the sill. Make sure that the threshold is horizontal. Insert wood shims between the jamb and the wall framing to keep the jamb square.
7. At each shimmed point, drill a counterbored hole through the jamb. Drive a wood screw into each hole and glue the wood plugs in place.
8. Tighten the hinges and make sure the door is still squarely aligned in the jamb.
9. Use a utility knife to cut off any excess shim from the edge of the door frame. Then run a continuous bead of caulk along the gap.
10. Nail the interior trim back around the door, and install any exterior trim. Run a continuous bead of caulk along the joint between the sill and the threshold, and between any siding and exterior trim.
11. Install locks on the door. (Chapter 3 explains how to choose and install locks.)

the door, they make the door more vulnerable to attack.

You can buy a door, side jambs, trim, threshold, and sill as separate parts, or you can get all the parts together in a single package—a door kit with precut jambs and sills, which is easier to install. You can also buy a door prehung (or preassembled), ready to be fastened to the rough opening.

Many modern door units come with sidelight panels (small vertical windows along the sides). If the sidelights might allow a burglar to climb through or to reach in for the lock, they should be made of or lined with a break-resistant material such as plastic.

How Strong Are Your Doors?

Door types ranked by resistance to attacks (listed in descending order):

- Flat steel, 0.3 inch or thicker
- Flat steel, between 0.1 inch and 0.3 inch thick
- Hardwood, 1.8 inch thick with solid wood panels
- Softwood, 1.8 inch thick with solid wood panels
- Softwood, 2 inches thick with glazed panels

SECURING YOUR WINDOWS

People who do a lot to secure their doors may be paying little attention to their windows because they think securing windows is time-consuming, expensive, or impossible. To burglars, windows are often the most attractive entry points.

The materials used in making doors are also used for manufacturing window frames. Wood, aluminum, fiberglass, and polyvinyl chloride (PVC) plastic are most popular for windows. As long as the windows are well-built and have good locking devices (keyless types are best), the frame material usually has little effect on a home's security.

Contrary to popular opinion, it usually isn't necessary to make your window frames and panes unbreakable to keep burglars out—unless your neighbors are out of earshot. Burglars know that few things attract more attention than the sound of breaking glass, and they don't like to climb through openings that have large jagged shards of glass pointing at them. When they can't get into a house without breaking a window, most burglars will move on to another house.

You can make your windows more secure just by making them hard to open quietly from outside. Don't install a lock or any other device that might delay a quick exit in case of a fire. Balancing the safety and security elements depends on what type of windows you have. The four basic types are: sliding, casement, louvered, and double hung.

A **sliding window** works much like a sliding glass door, and, like a sliding glass door, it usually comes with a weak lock that's easy to defeat. Most of the supplemental locking devices available for sliding windows fit along the track rail and are secured with a thumbscrew. (See Figure 2–10.) You can then keep the window in a closed or a ventilating position, depending on where you place the thumbscrew. The need

Figure 2–10. You can use thumbscrew devices to secure a sliding glass window. (Courtesy of VSI Donner)

to twist a thumbscrew can be inconvenient if you must frequently lock and unlock a window.

Another keyless device for securing sliding windows is offered by M.A.G. Enginnering & Manufacturing, Inc. (see Appendix B). The company's "Quick-Vent" model 8830 is a heavy-gauge-metal locking device that allows you to secure a window in three different positions. The device is held in place by security set screws instead of thumbscrews. Quick-Vent can be purchased for less than $10.

A **casement window** is hinged on one side and swings outward (much like your doors do). It uses a crank or handle for opening and closing (see Figure 2–11). To prevent someone from breaking the glass and turning the crank, the handle should be removed when it isn't being used.

Louvered (or "jalousie") windows are the most vulnerable type. They are made of a ladder-like configuration of narrow, overlapping slats of glass that can easily be pulled out of the thin metal channels. Jalousies attract the attention of burglars and should be replaced with another type of window.

The type of window used in most homes is the **double-hung window** (see Figure 2–12). It consists of two square or rectangular sashes that slide up and down and are secured with a metal thumb-turn butterfly sash "lock." (Although most manufacturers call it a lock, the device is really just a clamp.) The device holds the sashes together in the closed position, but a burglar can work it open by shoving a knife in the crack between the frames.

Figure 2–11. Casement windows swing open like doors. (Courtesy of Marvin Windows & Doors)

Figure 2–12. The double-hung window is the type most commonly used in homes. (Courtesy of Marvin Windows & Doors)

Watchguard Inc. (see Appendix B) makes a useful replacement for conventional sash locks. The company's Safety Sash Lock can't be opened from the outside of a home. It looks like a standard sash lock but incorporates a spring-loaded lever that prevents it from being manipulated out of the locked position. The Safety Sash Lock sells for about $10.

As an alternative to replacing sash locks, a ventilating wood window lock can be installed. (See Figure 2–13.) This device allows someone inside to raise the window a few inches and then set the bolt, which prevents anyone outside from raising it higher. It consists of an L-shaped metal bolt assembly and a small metal base. The bolt assembly fits along the inner edge of either of the two stiles (vertical members) of the top sash and is held in place with two small screws. The base is attached, in alignment, on the top of the bottom sash and is held in place with one small screw.

The bolt assembly has a horizontal channel that allows you to slide the bolt into the locked and unlocked positions. When in the locked position, the bolt extends over the base and obstructs the bottom sash from being raised past the bolt. When in the unlocked position, the bolt is parallel to the window and out of the way of the bottom sash. The higher you place the bolt mechanism above the bottom sash, the higher you'll be able to raise the window with the bolt in the locked position. The base isn't really needed, but it helps prevent the bottom sash from getting marred.

Many companies make ventilating wood window locks, and there aren't any important differences between brands. Most models are sold at home improvement centers and hardware stores for less than $5 each.

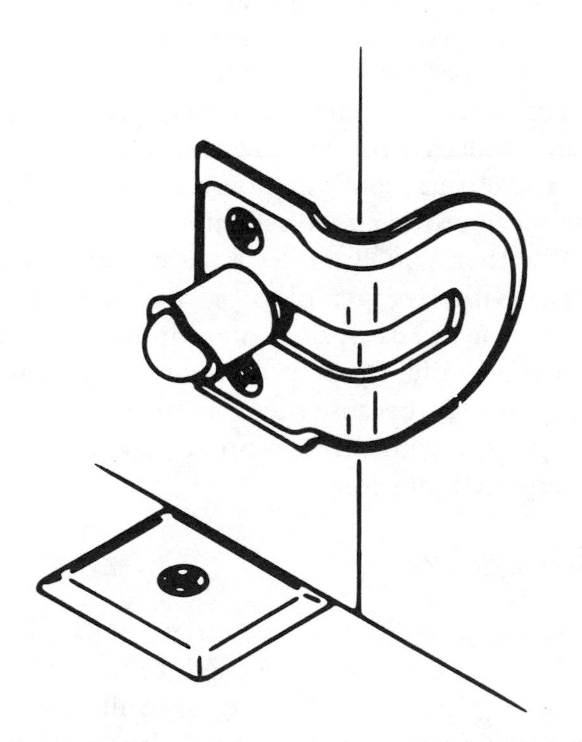

Figure 2–13. A ventilating wood window lock prevents a double-hung window from being raised too far. (Courtesy of VSI Donner)

Glazing

Glazing is a term that refers to any transparent or translucent material—usually some kind of glass or plastic—used on windows or doors to let in light. Most types of windows can be made more secure by replacing the glazing with more break-resistant material.

The most common glazing for small windows is standard sheet glass. It's inexpensive but it can easily splinter into small, sharp pieces. Plate glass, which is a little stronger than sheet glass, is generally used in large picture windows. Because plate glass also has the problem of breaking into many dangerous pieces, it shouldn't be used in exterior doors.

Tempered glass is several times stronger than plate glass and costs about twice as much. Rather than shattering into many sharp pieces, tempered glass breaks into small harmless pieces—the reason for using it in patio doors.

How to Secure a Double-Hung Window at No Cost

1. From inside the home, close the window and clamp the butterfly twist-turn sash lock into the closed position.
2. Use a pencil to mark two spots below the twist-turn lock on the top rail (horizontal member) of the bottom sash. One mark should be about 1 inch inside the left stile; the other should be about 1 inch inside the right stile.
3. Position your drill at the first mark and drill at a slightly downward angle until your drill bit goes completely through the top rail of the bottom sash and about halfway through the bottom rail of the top sash. Then do the same thing at the other mark you made.
4. Raise the bottom sash about 5 inches and hold it steady. Insert the drill bit back into one of the bottom sash holes and drill another hole about halfway through the top sash. (The hole should be about 5 inches above another hole you drilled on the stile.) Without moving the sash, do the same thing at the other side of the window.
5. Close the window and insert two small nails or eye bolts into the lower sets of holes to hold the sashes together so the window can't be lifted open from outside. When you want ventilation, you can remove the nails or bolts, raise the window, and insert them in the top set of holes to secure the window in the open position.

When a large piece of tempered glass breaks, it makes a lot of noise, which may attract the attention of neighbors.

The strongest type of glass a homeowner might use is laminated glass: It's made of two or more sheets of glass with a plastic inner layer sandwiched between them. The more layers of glass and plastic, the stronger (and the more costly) the laminated glass will be. Laminated glass 4 inches thick can stop bullets and is often used for commercial applications.

Plastics are commonly used as glazing materials. Acrylics such as "Plexiglas" and "Lucite" are very popular because they are clearer and stronger than sheet glass. However, they scratch easily and can be sawed through.

The strongest types of plastic a homeowner might use are polycarbonates like "Lexigard" and "Lexan." Although they're not as clear as acrylics, polycarbonates are up to 30 times stronger. Untreated polycarbonates scratch easily, but you can buy sheets with scratch-resistant coatings.

For increased strength and energy efficiency, many modern windows come in parallel double- or triple-pane configurations. Three parallel panes can provide good security.

Whether you want to replace your glass panes with stronger glass or replace broken panes, you can easily do it yourself. In addition to being unsightly, a broken window can attract burglars because they can quietly remove the glass to gain entry. A broken pane should be replaced immediately.

Glass Blocks

Glass blocks come in a wide variety of patterns and sizes, and are strong enough to be used in place of plate glass. They're especially useful for securing basement windows. Most patterns create a distorted image to anyone trying to see through them, but clear glass blocks are also available.

How to Replace a Glass Pane

1. Working from the exterior side of the window, use a wood chisel to remove any soft or crumbling glazing compound from along the channel between the glass and window pane. Soften any remaining unpainted glazing compound by applying a heavy coat of linseed oil and waiting about 30 minutes. Then remove all the unpainted glazing compound. Soften any remaining painted glazing compound by using a heat gun, and remove the rest of the glazing compound.
2. After all the glazing compound has been removed, use a putty knife or long-nose pliers to work the glazier's points (small clips used to secure the pane) away from the window pane. Then lift the pane of glass out of the window.
3. Use a wire brush and medium sandpaper to clean and smooth out each channel. (You may then want to paint the channels for aesthetic reasons.)
4. Insert your new pane of glass or plastic. Seal the pane with whatever type of glazing compound the manufacturer recommends. Peel the paper from the new pane and clean the pane with warm water and a mild detergent.

For areas that require ventilation, you can buy preassembled panels of glass block with built-in openings. Preassembled panels are easy to install if you get the right sizes. To order the right size panel, you need to know the size of your window's rough opening. If you have a wood-frame wall, you can determine the rough opening by measuring the width of the opening between the frame's sides and the height between the sill and the header. You'll need a panel about 1/2 inch smaller than that measurement, to make sure it will fit in easily.

If you have a masonry wall, you can determine the rough opening by measuring the width between the brick or block sides and the height between the header and the sill. Be sure that the panel you order is about 1/2 inch smaller than the opening.

There are two basic ways to install glass block panels. The older way involves using masonry cement or mortar—in much the same way as when installing bricks. The newer way involves using plastic strips.

A Newer Way to Install Glass Blocks

A cleaner and simpler way to install glass blocks was recently developed by Pittsburgh Corning. However, the company's glass-block panel kits aren't designed to offer strong resistance to break-in attempts. The kit includes: U-shaped strips of plastic channel for the perimeter of the panel; a roll of clear plastic spacer for holding the blocks in place; clear silicon caulk; and a

How to Install Glass Blocks with Mortar

1. For a masonry wall installation, insert shims in the opening around the glass block panel. Make sure the panel is squarely in place.
2. Use a trowel to push mason's mortar into the gaps around the perimeter of the panel.
3. After the mortar sets, remove the shims and fill their spaces with mortar. Seal the panel with a continuous bead of caulking.

joint-cleaning tool. The kit can be used in the following way:

1. Install the U-shaped channels, using shims if necessary. Attach the channels with 1-inch flathead wood screws, and conceal the screw heads with white paint.
2. Using short lengths of spacer, align the blocks vertically. (Use a utility knife to cut spacers to size.)
3. Align the blocks horizontally with long strips of spacer. (The spacer's contour matches the block edge surface.)
4. Fit the last block into the panel through a section temporarily removed from the top channel.
5. Slip the leftover section of the top channel over the last block, and apply caulk to hold the section in place.
6. Wipe the joints clean with a cloth dampened with isopropyl alcohol. Fill the joints with beads of silicon caulk.

Protecting Glass

If you're concerned that someone will gain entry by breaking a window, you might want to coat it with security film, a transparent laminated coating that resists penetration and firmly holds broken glass in place. (See Figure 2–14.) Even after breaking the glass, burglars would have a hard time getting through. Some security film can hold glass in place against sledgehammer blows, high winds, and explosions.

A less aesthetic but equally effective way to protect windows is to install iron security bars. The mounting bolts should be reachable only from inside your home. Be sure the bars don't make it hard for you to escape quickly if necessary. (See Figure 2–15.) Hinge kits are available for many window bars.

Before buying window bars, you'll need to measure the width and height of the area to be covered. In general, the larger the area to be protected, the more the bars will cost. Window bars range in price from about $10 to more than $50 per window. (See Figure 2–16.)

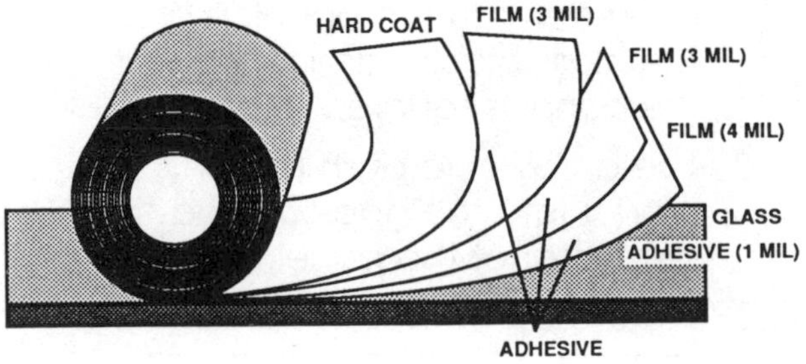

HARD COAT
FILM (3 MIL)
FILM (3 MIL)
FILM (4 MIL)
GLASS
ADHESIVE (1 MIL)
ADHESIVE

Figure 2–14. Security film is a laminated coating that protects glass. (Courtesy of CHB Industries)

Figure 2–15. Bars can be a good way to protect a window if they are designed to allow you to use the window as an emergency exit. (Courtesy of Leslie-Locke, Inc.)

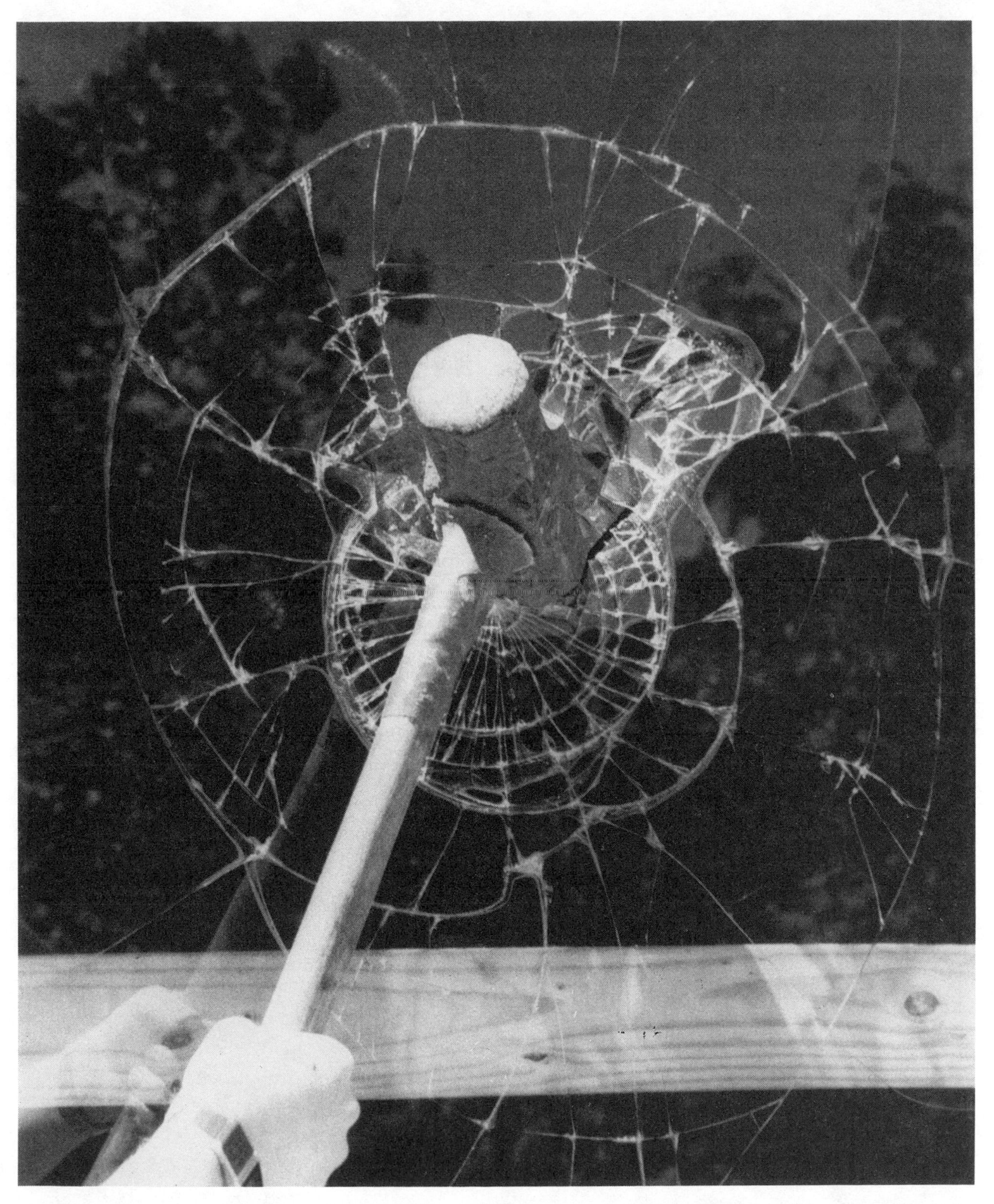

Figure 2–16. Some types of security film can make a window resist blows from a sledgehammer. (Courtesy of CHB Industries)

RECOMMENDED RESOURCES

(See HM Sourcelist for more information.)

Andersen Windows—Doors, windows

Atrium Door & Windows Company—Doors, windows

Benchmark, General Products—Doors

Bennett Industries, Inc.—Doors

Bilco Company—Doors

Brookfield Industries, Inc.—Hinges

Caradco—Doors, windows

CHB Industries—Security film

Cole Sewell Corporation—Doors, windows

Crestline—Doors

Don-Jo Manufacturing—Strike boxes

Jeld-Wen, Inc.—Doors, windows

Jessup Door Company—Doors

Johnson Products, Inc.—Doors

Latch-Gard—Door hardware

Leslie-Locke, Inc.—Bars, doors, grilles, windows

M.A.G. Engineering & Manufacturing Company, Inc.—Strike boxes, window locks

The Malta Company—Doors, windows

Markar Products, Inc.—Hinges

Marvin Windows & Doors—Doors, windows

Meister Atlanta Corporation—Strike boxes

Morgan Manufacturing—Doors

MSI Mace—Door braces, window locks

Nana Windows & Doors—Doors, windows

Norco Windows, Inc.—Doors, windows

Peachtree Windows & Doors—Doors, windows

Pease Industries—Doors

Pella Rollscreen Company—Doors, windows

Perma-Door Mfg.—Doors

Pinecrest Inc.—Door and window hardware, doors, windows

Pittsburgh Corning Corporation—Glass blocks

Rudolph-Desco Company, Inc.—Door viewers

Simpson Door Company—Doors

Stanley Door Systems—Doors

Stanley Hardware—Door locks, hinges, window locks

Stor-A-Dor—Doors

Taylor Brothers—Doors

Therma-Tru Corporation—Doors

Velux-America, Inc.—Windows

Vigilante Burglar Bars—Bars, grilles

VSI Donner—Door locks, window locks

Watchguard Inc.—Door locks, window locks

Weather Shield Mfg.—Doors, windows

Wenco Windows—Windows

3

Selecting a Good Lock

If you have strong doors and door frames, you'll need strong locks to secure them against break-in attempts. (If you've never thought about your doors and door frames, Chapter 2 explains how important they are.)

This chapter shows how burglars defeat locks, how to choose the best locks for your particular needs, how to make the most of the locks you have, and how to install some of the new types of locks.

GRADES OF DOOR LOCKS

Most lock manufacturers offer locks in several grades, referred to as "light duty," "residential," "heavy duty," and "commercial." The grade descriptions are useful guidelines, but there are no industry standards for manufacturing each grade. One manufacturer's "heavy duty," for example, may be of lower quality and less resistant to break-ins than another manufacturer's "light duty." The grade names are meaningful only for comparing locks made by the same manufacturer.

A better measure of a lock's quality is the rating given to it by the American National Standards Institute (ANSI). The three most common ANSI standards for locks—Grade 1, Grade 2, and Grade 3—are based on a range of performance features, especially during impact tests. The ANSI tests gauge how well the locks resist forceful attacks and whether their finishes hold up well, over extended periods of time.

When you see one of these ANSI ratings, this is what you should know:

Grade 1 locks are for heavy-duty commercial uses and would be overkill for most homes.

Grade 2 locks, although designed for light commercial uses, can provide the best protection for most homes.

Grade 3 locks are good for light residential applications.

Most door locks sold to homeowners are either Grade 3 or have no ANSI classification.

Only a few lock manufacturers—Kwikset Corporation and Master Lock Company, for example—offer lines of Grade 2 locks through department stores and home improvement centers. These ANSI-graded locks often cost little more than locks that have no ANSI classification.

Should you purchase a lock that doesn't have an ANSI classification? Aside from the directions on how to install it, you can't rely on any information the manufacturer has printed on the packaging. Much of the description of what the lock can do for you is just advertising hype. To be able to separate the hype from meaningful information, you'll need to know the strengths and weaknesses of the generic types of door locks.

TYPES OF DOOR LOCKS

Six types of locks are commonly used to secure doors in homes:

1. Bit-key
2. Key-in-knob
3. Key-in-lever
4. Deadbolt
5. Jimmyproof deadlock
6. Multiple bolt

You may not be familiar with their names, but you've probably seen all six types of locks. The first three are not recommended as sole exterior door locks.

The bit-key, the oldest type, has a large keyhole and works with a "*skeleton key.*" A bit-key lock is all right for closet or bathroom doors, but shouldn't be used on exterior doors because it's very easy to defeat. Many people know how to unlock a bit-key with a metal coat hanger, and anyone who buys the stock bit-keys sold at many hardware stores can probably unlock almost any bit-key lock. Because there isn't a lot of variation among bit-keys, someone who buys ten different bit-keys could use them to unlock most bit-key locks in homes throughout the United States!

The most popular exterior door lock installed in homes built since the late 1950s is the key-in-knob. Basically, the key-in-knob is two connecting doorknobs with a keyway in one doorknob or in both (see Figure 3–1). The key-in-knob is inexpensive and easy to install. It allows the door to be locked or latched—without a key—just by closing the door.

Like a bit-key lock, a key-in-knob is easy to defeat. It can't be opened with over-the-counter keys or coat hangers, but the key-in-knob is vulnerable in other ways. After hammering one of the knobs off, for instance, a burglar can use a screwdriver to retract the lock's bolt from the door frame. There are also quieter ways to defeat the key-in-knob lock. Like the bit-key lock, the key-in-knob is fine for a bathroom or closet door but shouldn't be used as the sole lock for an exterior door.

Internally, the key-in-lever is designed a lot like the key-in-knob. Both have the same vulnerabilities. The key-in-lever consists of one or two levers instead of knobs. The levers make it easier for young children and people with physical handicaps to use these locks.

If your door has one of the three types of locks just described, it may be convenient for you to gain extra security by installing a stronger lock above your present key-in-lever, key-in-knob, or bit-key lock. You might want to use a deadbolt, jimmyproof deadlock, or multiple-bolt lock. Each is described and illustrated in the sections that follow.

Figure 3–1. Although the key-in-knob is popular among homeowners, it provides little security when it's the sole lock on an exterior door. (Courtesy of VSI Donner)

Choosing a Deadbolt

Many police departments recommend using a tubular deadbolt (or "deadbolt," for short) (see Figure 3–2) on exterior doors. However, some deadbolts aren't much better than a key-in-knob lock. To understand why, you have to know how deadbolts work and how burglars try to defeat them.

The main parts of a typical deadbolt include: a cylinder, cylinder guard, bolt assembly, tailpiece, thumb-turn, and mounting screws (see Figure 3–3). The cylinder is the cylindrical metal part with a keyway on its face, and the cylinder guard (or "collar") fits around the cylinder. Both fit together and are mounted on the exterior side of the door. Using the mounting screws, the thumb-turn is mounted behind the cylinder on the interior side of the door. The bolt assembly and tailpiece fit within a cavity in the door, between the thumb-turn and cylinder. It is the tailpiece that connects the thumb-turn and cylinder to the bolt assembly.

Here's how a deadbolt works. On the exterior side of the door, when you turn a key in the cylinder, the connecting tailpiece engages the bolt mechanism, which causes the bolt to extend or retract. On the interior side, when you twist the thumb-turn, the connecting tailpiece engages the bolt mechanism and causes the bolt to extend or retract.

Some high-security deadbolts have special parts. The Abloy Disklock, for example, includes adapter rings and a bolt assembly protector. The bolt assembly protector fits between the cylinder and the bolt assembly and has a

Figure 3–2. When installed on a strong door, a deadbolt lock and high-security strike box can provide great security. (Courtesy of Abloy Security, Inc.)

circular opening to allow the tailpiece to be connected to the cylinder. Because it has a hood that covers the internal parts of the bolt assembly, the bolt assembly protector prevents a burglar from manipulating the bolt assembly with an ice pick or other instrument. Adapter rings are used to make a lock fit better on a thin door. They are installed between the door and the cylinder guards.

How Deadbolts Are Defeated and What You Can Do about It

Common methods of attacking deadbolts include jimmying, sawing, wrenching, and lockpicking.

Jimmying. Jimmying is done by inserting a pry bar between a door and its frame, near the extended bolt, and prying until the bolt is freed from its strike place. The longer the bolt, the harder the lock will be to jimmy. Make sure your deadbolt has a 1-inch throw: one inch of the bolt should extend past the edge of the door.

Sawing. If the bolt is made of a soft metal, such as brass, it can be sawed off with a hacksaw blade. Make sure your lock's bolt is made of steel or has a hardened steel insert.

Wrenching. When a wrench is clamped onto a cylinder guard, the cylinder can be twisted off of the door. Only use deadbolts that come with tapered free-spinning cylinder guards. They're hard to wrench because they just spin around the cylinder rather than turning it.

Lockpicking. In theory, any lock that uses a key can be picked. It's true that it's always *possible* to create an instrument that can be used to simulate the action of a key, but most people don't need to worry that a burglar will pick-open their locks. Contrary to what's portrayed in movies, very few home burglars pick-open door locks. Lockpicking is a sophisticated skill that takes a long time to learn, and there are usually faster and easier ways to break into a home.

ABLOY DISKLOCK® 2200 SERIES
Single Cylinder Deadbolt

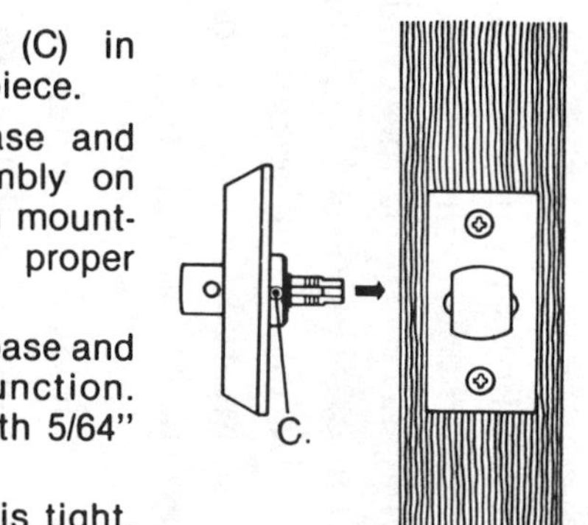

1. Insert thumb-turn tailpiece into thumb-turn assembly socket (A). Make sure the adjusting ring (B) is on the tailpiece.

2. With deadbolt in extended position and the flat side of the tailpiece facing the edge of the door, push thumb-turn assembly into place while supporting the bolt from the opposite side. This will adjust the length of the tailpiece to the door thickness.

3. Tighten set screw (C) in socket to secure tailpiece.
4. Place thumb-turn base and outer cylinder assembly on the door. Fasten with mounting screws cut to proper length
5. Place thumb-turn on base and test for proper function. Tighten set screw with 5/64" (2mm) hex wrench.

   NOTE: If operation is tight, shorten tailpiece slightly.

Figure 3–3. Most parts of a deadbolt lock can't be seen when it's installed on a door. (Courtesy of Abloy Security, Inc.)

Suppose you lost your last set of house keys and chose not to break a window or glass panel to gain entry into your home. An experienced locksmith may need 10 or more minutes to pick-open your exterior door lock. This professional would be intent on opening the door without damaging the lock or the door. Burglars are concerned about gaining entry quickly and they care little about the damage they might cause. Why would burglars spend 5 to 10 frustrating minutes fiddling with lockpicking when they can kick the door in or jimmy the lock instead?

If you live in a large city or near a high crime area, however, you may need to be concerned about lockpicking. Such places have more than their share of sophisticated burglars who can pick-open locks faster than many locksmiths can. The best way to thwart lockpicking attempts is to use high-security cylinders in the locks on all of your exterior doors. (Information about high-security cylinders is given later in this chapter.)

Installing a Deadbolt Lock

A deadbolt lock is installed by drilling two holes: a larger hole through the face and back of the door, and a smaller hole through the edge of the door. The larger hole is usually either 2 1/8 inches or 2 3/8 inches in diameter. The smaller hole is usually 1 inch in diameter or smaller. The distance from the door edge to the center of the large hole is called the *backset.* Most backsets are either 2 1/2 inches or 2 3/4 inches. (See Figure 3–4.)

You need to know your door's exact measured distances before you buy or install any door lock. For a new installation, you need the measurements to choose the right size drill bits. For a retrofit installation, you need the measurements to choose a lock that lets you use the pre-existing installation holes.

Most deadbolts come with a template. Wrap the template around the edge of the door and mark on the door the spots where you are to drill the holes. Be sure the template is straight, and place your marks on the high side of the door bevel. (The lock edge of your door is slightly beveled so the door can open and close easily.)

When drilling a lock hole through a door, always use a pilot drill bit that extends at least 1 inch past your hole saw. Begin drilling on the high side of the door. When your pilot bit penetrates through the other side of the door, stop drilling. Go to the side of the door that your pilot bit punctured, and use the small puncture hole as a guide to finish drilling the lock hole. By working on both sides of the door when you're drilling a lock hole, you'll be less likely to splinter the door.

Be sure to *drill straight.* If you have a hard time drilling straight, you might want to attach a small level onto your drill or use a lock hole boring jig.

If you're installing a deadbolt on a metal door or a hollow-core wood door, you'll need to install a lock support insert in the door to prevent the lock from loosening. Lock support inserts generally cost less than $10.

Choosing a Jimmyproof Deadlock

A jimmyproof deadlock (or "vertical deadbolt") can offer a lot of resistance to jimmying attempts. It has a rectangular body with two or three cylindrical bolts at one end (see Figure 3–5). The lock is surface-mounted on a door so that its bolts are aligned with the "eye loops" of a matching strike plate. When placed in the locked position, the bolts drop vertically into the eye loops, and the lock can't be separated from the strike plate without breaking the entire unit.

One problem with a jimmyproof deadlock is that it's effective only when properly installed on a strong door frame and door. In many

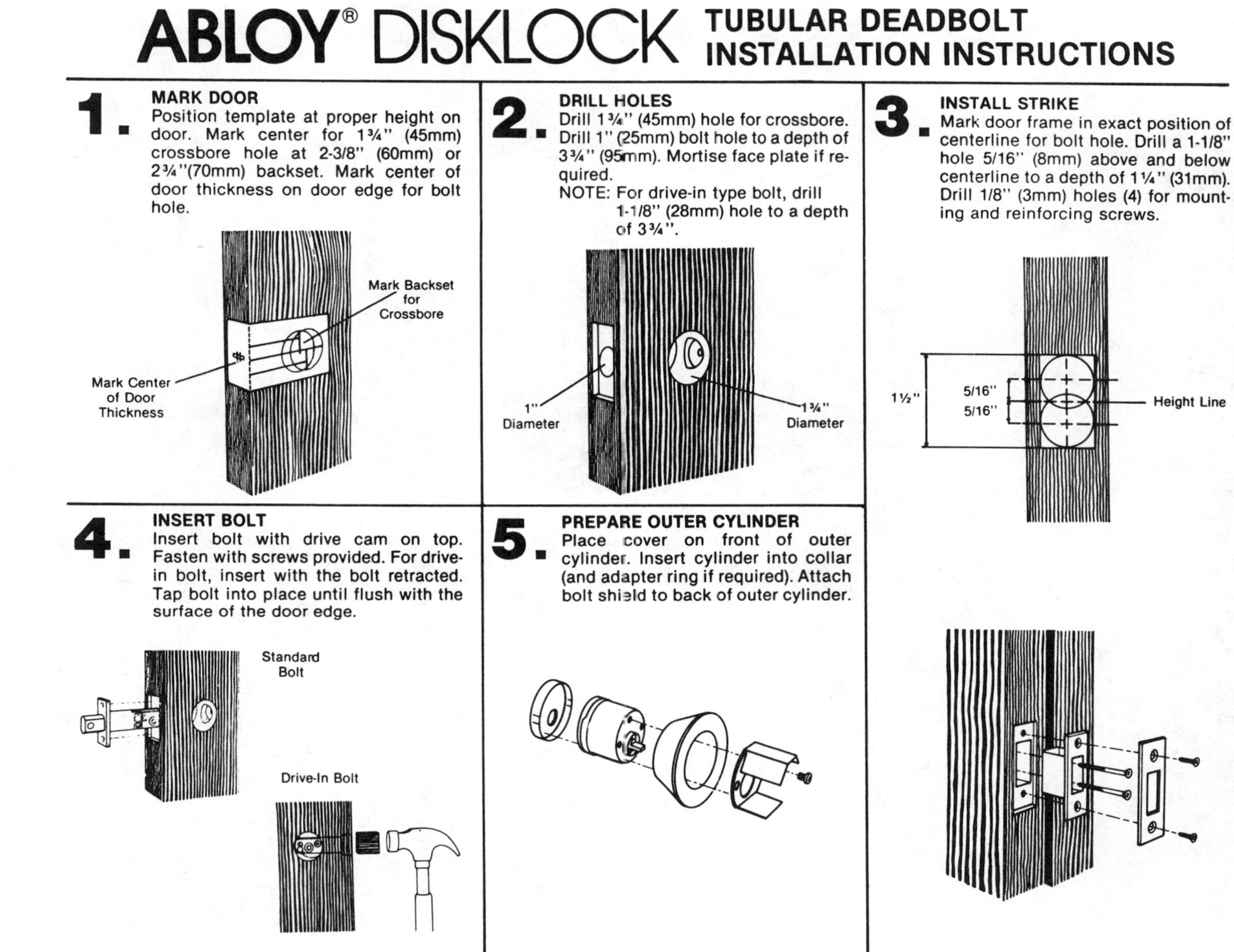

Figure 3–4. Although they may require different hole sizes and drilling positions, most deadbolt locks are installed in basically the same way. (Courtesy of Abloy Security, Inc.)

Figure 3–5. A jimmyproof deadlock is designed to offer special resistance to jimmying attempts. (Courtesy of Medeco Security Locks)

homes, the door frames aren't strong enough to support a jimmyproof deadlock properly. Another problem is that if the mounting screws for the strike plate aren't long enough, the strike plate won't hold to the frame during a kick-in. Be sure to use screws that are at least 3 inches long when you install a strike plate.

In jimmyproof deadlocks, unlike deadbolts, the cylinders aren't protected by cylinder guards, which makes jimmyproof deadlocks vulnerable to drilling, pulling, and chiseling attacks. (These sophisticated types of attacks rarely occur outside of large cities and high crime areas.) The easiest way to thwart these attacks is to install a hardened steel cylinder guardplate over the jimmyproof deadlock. (See Figure 3–6.) The guardplate covers virtually every vulnerable area of the cylinder, leaving only the keyway exposed. (See Figures 3–7 and 3–8.)

Figure 3–6. In high-crime areas, a cylinder guardplate should be installed with a jimmyproof deadlock. (Courtesy of Medeco Security Locks)

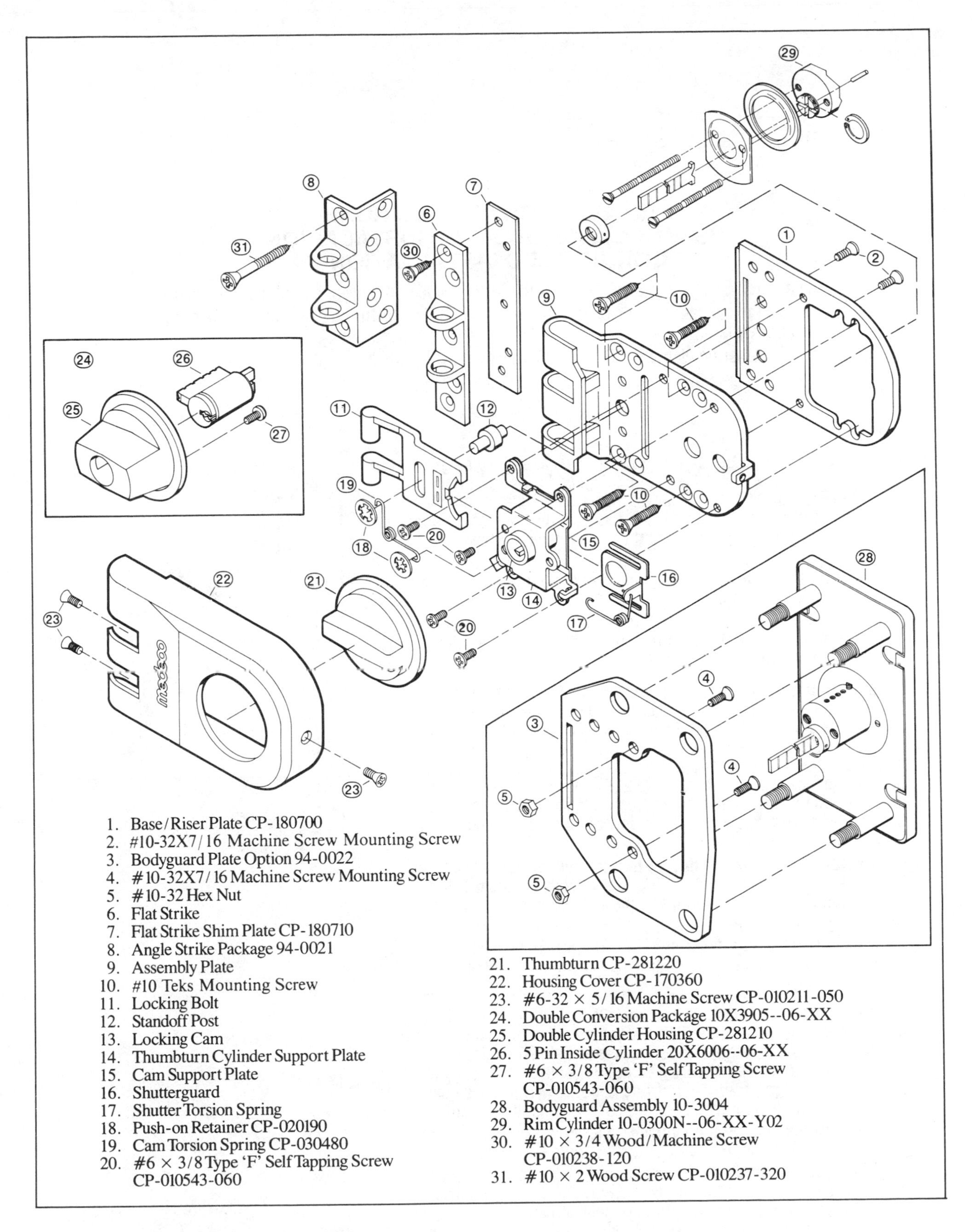

Figure 3–7. The Medeco Bodyguard is a popular cylinder guardplate designed especially for Medeco brand jimmyproof deadlocks. (Courtesy of Medeco Security Locks)

STEELGUARD

INSTALLATION INSTRUCTIONS
INSTRUCCIONES PARA SU INSTALACION
CONSIGNES D'INSTALLATION

The following are instructions for the STEELGUARD with a Medeco rim cylinder and the Bodyguard.

A continuación se indican las instrucciones para instalar el STEELGUARD en cerraduras de cilindro de Medeco con el Bodyguard.

Voici comment installer un élément STEELGUARD avec un barillet Medeco et l'élément Bodyguard.

1

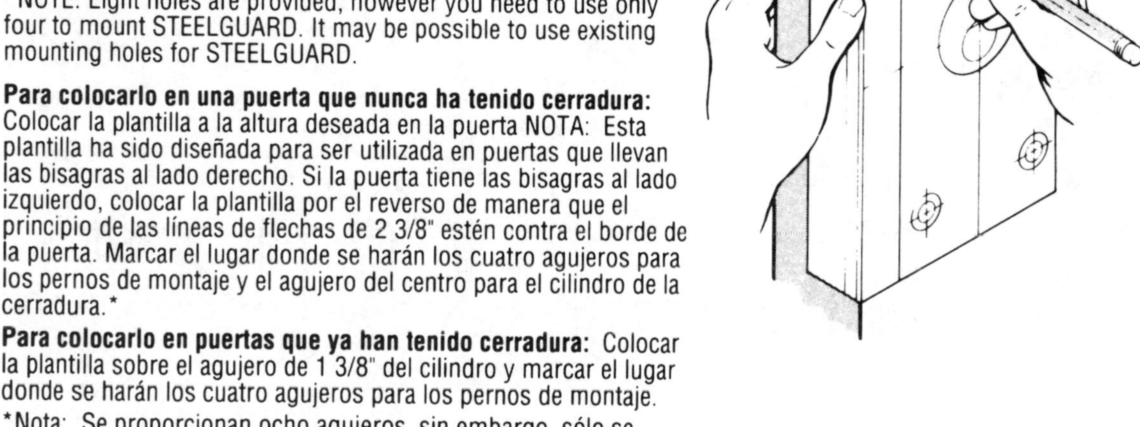

For new door applications: Position template at desired height on door. NOTE: This template has been designed for use on right-hinged doors; for left-hinged doors, reverse template so that the beginning of the 2 3/8" arrow lines up against the edge of the door. Mark the four mounting hole, and the center holes for the lock cylinder.*

For existing doors: Position template over 1 3/8" cylinder hole, and mark four mounting holes.

*NOTE: Eight holes are provided, however you need to use only four to mount STEELGUARD. It may be possible to use existing mounting holes for STEELGUARD.

Para colocarlo en una puerta que nunca ha tenido cerradura: Colocar la plantilla a la altura deseada en la puerta NOTA: Esta plantilla ha sido diseñada para ser utilizada en puertas que llevan las bisagras al lado derecho. Si la puerta tiene las bisagras al lado izquierdo, colocar la plantilla por el reverso de manera que el principio de las líneas de flechas de 2 3/8" estén contra el borde de la puerta. Marcar el lugar donde se harán los cuatro agujeros para los pernos de montaje y el agujero del centro para el cilindro de la cerradura.*

Para colocarlo en puertas que ya han tenido cerradura: Colocar la plantilla sobre el agujero de 1 3/8" del cilindro y marcar el lugar donde se harán los cuatro agujeros para los pernos de montaje.

*Nota: Se proporcionan ocho agujeros, sin embargo, sólo se necesitan cuatro para montar el STEELGUARD. Es posible que se puedan utilizar los agujeros de montaje anteriores para colocar el STEELGUARD.

Pose sur portes sans trou de verrou : placez le gabarit à la hauteur désirée sur la porte. REMARQUE : le gabarit est prévu pour la pose sur portes s'ouvrant à gauche ; si votre porte s'ouvre à droite, retournez le gabarit de manière à ce que le début de la flèche de 2 3/8 po. soit aligné contre le bord de la porte. Marquez les quatre trous de montage et un trou central pour le barillet du verrou.*

Pose sur portes avec trou de verrou : alignez le gabarit sur le trou du barillet de 1 3/8 po. et marquez quatre trous de montage.

*REMARQUE : Il y a huit trous; cependant, vous n'avez besoin que d'en utiliser quatre pour monter le STEELGUARD. Vous pourrez peut-être utiliser les trous de montage déjà percés.

Figure 3–8. You can install a Medeco Bodyguard by following these ten steps. (Courtesy of Medeco Security Locks)

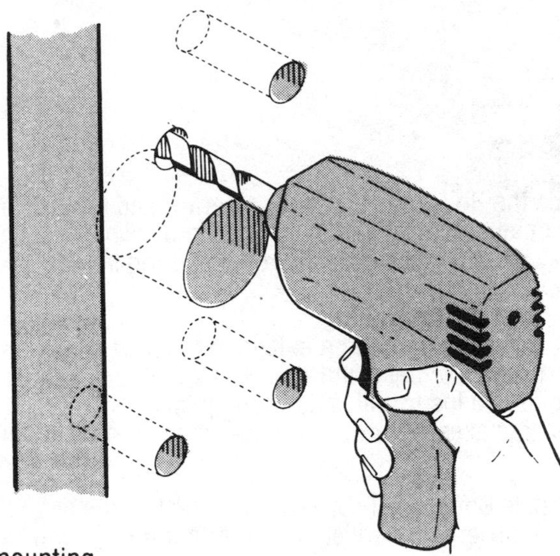

2

Drill 1 3/8" cylinder hole through the door; drill four mounting holes, 11/64" diameter and a minimun of 1/2" deep, for #10 screws.

Taladrar un agujero de 1 3/8" de diámetro en la puerta para colocar el cilindro; taladrar los cuatro agujeros para los pernos de montaje #10, los agujeros deben ser de 11/64" de diámetro y un mínimo de 1/2" de profundidad.

Percez au travers de la porte un trou de 1 3/8 po. de diamètre pour la cavité du barillet ; percez quatre trous de montage de 11/64 po. de diamètre et au moins 1/2 po. de profondeur pour les vis n° 10.

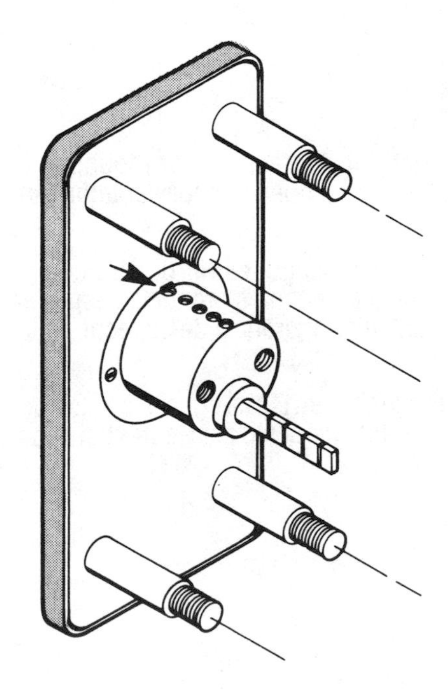

3

Anchor rim cylinder in Bodyguard with the steel cylinder retainer plate, making sure that cylinder tailpiece is in the vertical position. Use the longer set screw (provided) in chamber #1.

Fijar el cilindro en el Bodyguard usando la placa de acero de retención del cilindro. Controlar que la cola del cilindro esté en posición vertical. Se suministra un tornillo más largo para utilizarlo en la cámara #1. .

Ancrez le barillet dans l'élément Bodyguard en posant la plaque de retenue en acier et en vous assurant que la tige du barillet est bien en position verticale. Introduisez la vis sans tête longue (fournie) dans la chambre de goupille n° 1.

Figure 3–8. *Continued.*

4

Place the Bodyguard on door exterior, and install cylinder back plate on door interior. Adjust rim cylinder mounting screws to door thickness before securing into door.

Colocar el Bodyguard en el lado exterior de la puerta y la placa posterior del cilindro en el lado interior de la puerta. La longitud de los tornillos de montaje del cilindro debe ajustarse al espesor de la puerta antes de la instalación.

Placez le Bodyguard sur l'extérieur de la porte et posez la plaque arrière du barillet sur l'intérieur de la porte. La longueur des vis de montage du barillet doit être adaptée à l'épaisseur de la porte avant le montage.

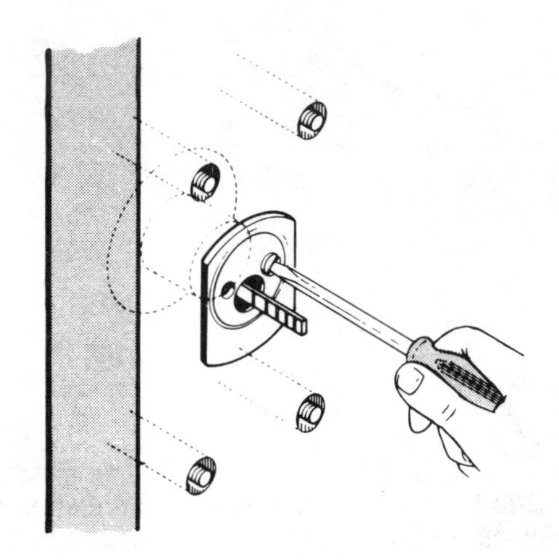

5

Slip Hardguard sleeves over mounting bolts and add spacer washers to compensate for variations in door thickness.

Colocar los manguitos Hardguard encima de los pernos de montaje y agregar las arandelas espaciadoras para compensar cualquier diferencia de espesor en la puerta.

Enfilez les renforts Hardguard sur les boulons de montage et ajoutez des bagues d'écartement si l'épaisseur de la porte le demande.

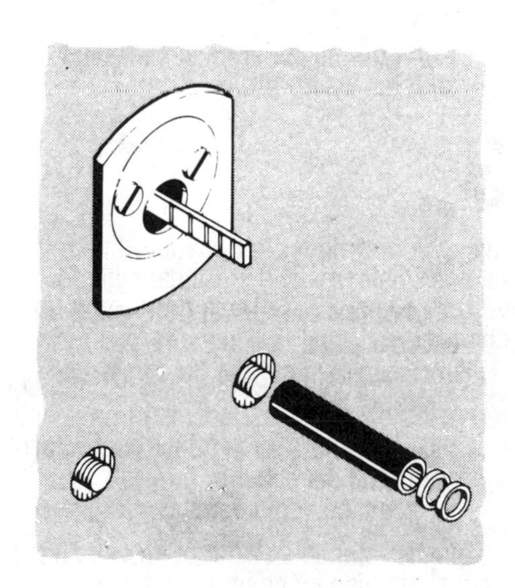

Figure 3–8. *Continued.*

6

Remove the STEELGUARD's base plate, held in place by two screws. Attach Bodyguard base plate with two screws, plus two additional screws using hex nuts.

Sacar la placa del STEELGUARD destornillando los dos tornillos que la sujetan. Colocar la placa del Bodyguard utilizando dos tornillos, más dos tornillos adicionales, utilizando tuercas hexagonales.

Retirez la plaque d'appui du STEELGUARD, laquelle est maintenue par deux vis. Fixez la plaque d'appui du Bodyguard à l'aide de deux vis, plus deux vis supplémentaires avec écrous hexagonaux.

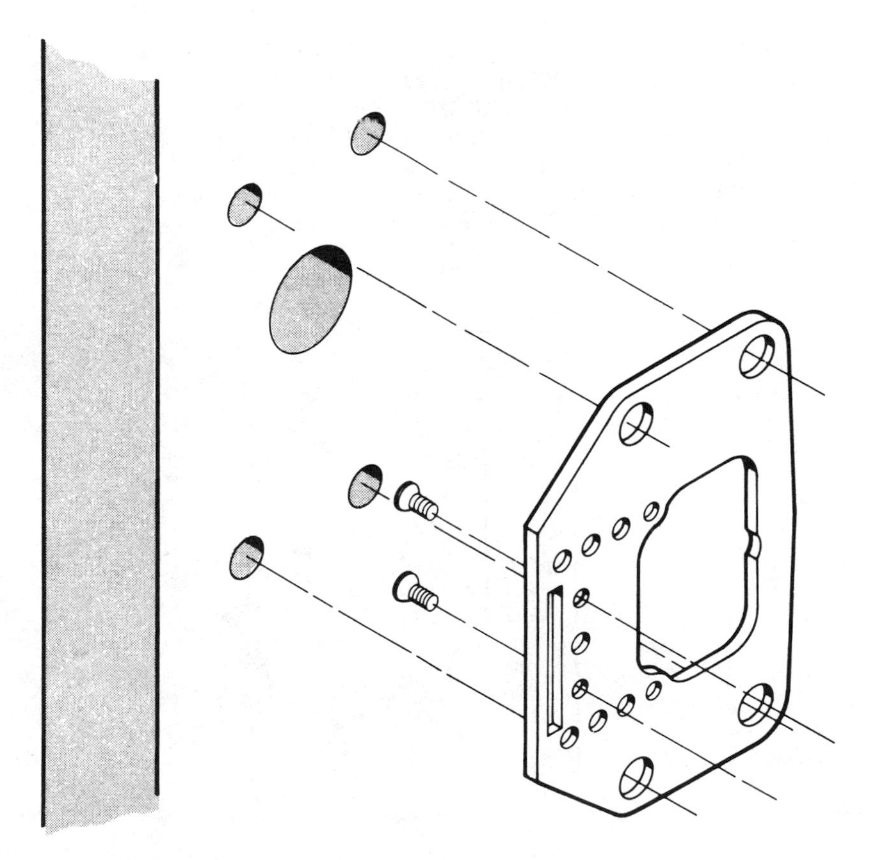

Figure 3–8. *Continued.*

7

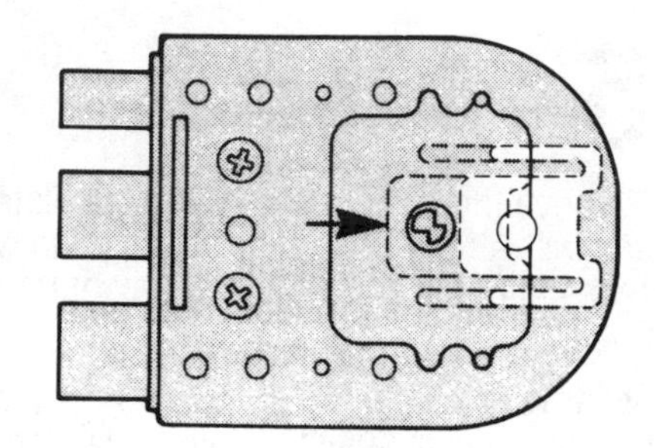

Slide the STEELGUARD's shutterguard horizontally so that the locking cam opening is revealed. Align rim cylinder tailpiece with this locking cam, and place STEELGUARD assembly plate on the door. Insert tamperproof sex bolts into mounting holes and onto four mounting bolts of Bodyguard. Use the slot drive adaptor to tighten assembly on door.

Colocar el shutterguard (postigo) del STEELGUARD en posición horizontal, de manera que aparezca la apertura de la leva del cerrojo. Alinear la cola del cilindro con este cerrojo y colocar la placa de ensamblaje del STEELGUARD sobre la puerta. Colocar los pernos hembra contra forzaduras en los agujeros de montaje y en los cuatro pernos de montaje del Bodyguard. Utilizar el adaptador para la punta del destornillador para apretar los pernos en la puerta.

Faites glisser la pièce Shutterguard du STEELGUARD latéralement de manière à dégager l'ouverture de la came. Alignez la tige du barillet avec cette came de verrouillage et posez la plaque de montage STEELGUARD sur la porte. Introduisez les boulons « femelles » anti-effraction dans les trous de montage et fixez-les sur les quatre boulons de montage du Bodyguard. Serrez l'ensemble sur la porte à l'aide de l'adaptateur de tournevis pour écrous à fente.

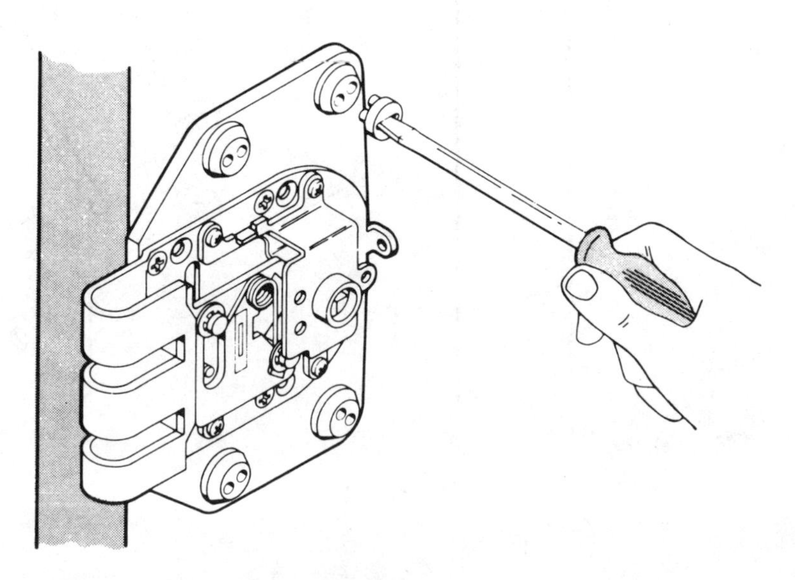

Figure 3–8. *Continued.*

8

For a finished appearance and tamperproof protection, tap the brass plugs into the holes on the sex blots.

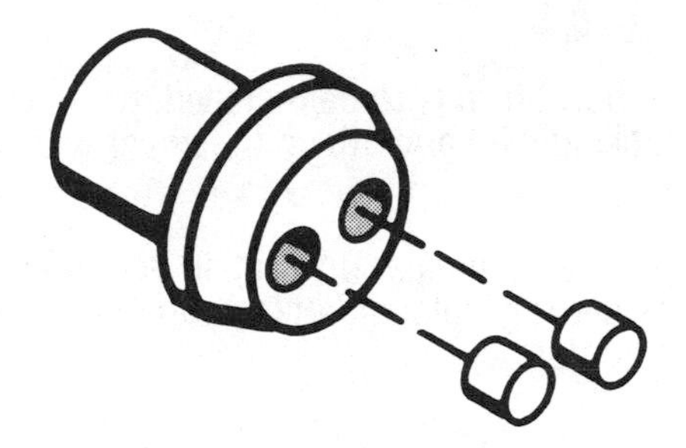

Para un mejor acabado o protección contra forzaduras, colocar los tacos de bronce en los agujeros de los pernos hembra.

Pour une finition nette et une sécurité accrue, enfoncez les protecteurs en laiton dans les têtes percées des boulons femelles.

9

Secure strike to door jamb, aligning its knuckles with the STEELGUARD's openings.

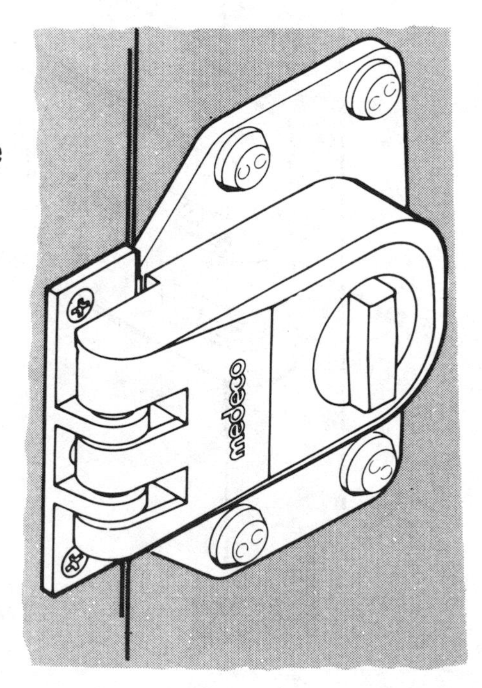

Colocar la hembra de la cerradura en el quicio de la puerta de manera que los nudillos queden alineados con las aperturas del STEELGUARD.

Posez la gâche sur le châssis de porte de manière à aligner les oeillets avec les ouvertures du STEELGUARD.

Figure 3–8. *Continued.*

10

If thumbturn is to be installed, position thumbturn onto locking cam and attach the cover with three enclosed screws.

Si se va a instalar un pestillo de botón, colocarlo sobre la leva de cierre y colocar la tapa con los tres pernos que se adjuntan.

Si vous comptez installer le bouton tournant, mettez-le sur la came de verrouillage et fixez le couvercle à l'aide des trois vis fournies.

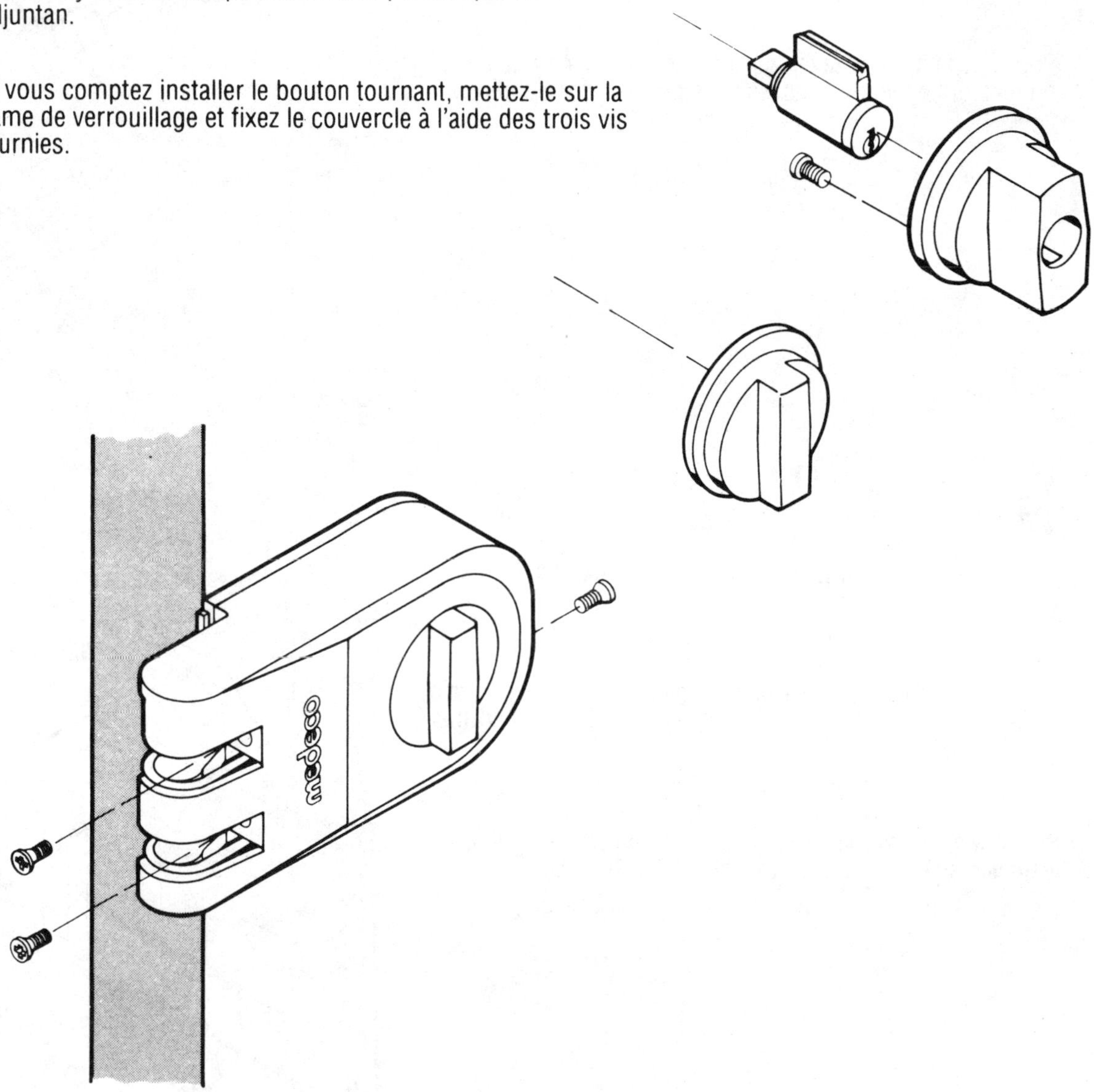

Figure 3–8. *Continued.*

Multiple-Bolt Locks

For maximum lock protection, you can use a multiple-bolt lock, a type of lock that has two or more long bolts that operate simultaneously. The bolts may extend vertically, horizontally, or in several directions. Usually, a key unlocks the door from the outside, and a thumb-turn moves the bolt from the inside. There are two basic kinds of multiple-bolt locks: surface-mount and mortise.

The surface-mount type has two or more bolts that extend and retract across a door's surface. A popular surface-mount style, often called a "police lock," has two bolts: one extends horizontally into the hinged side of a door and the other extends into the door's opening side. (See Figures 3–9 and 3–10.)

The mortise type of multiple-bolt lock has bolts that move within a hollowed-out cavity in the door. These locks can be used only on a thick wood or metal door. They usually need to be installed by a professional who has the special tools needed to prepare the door.

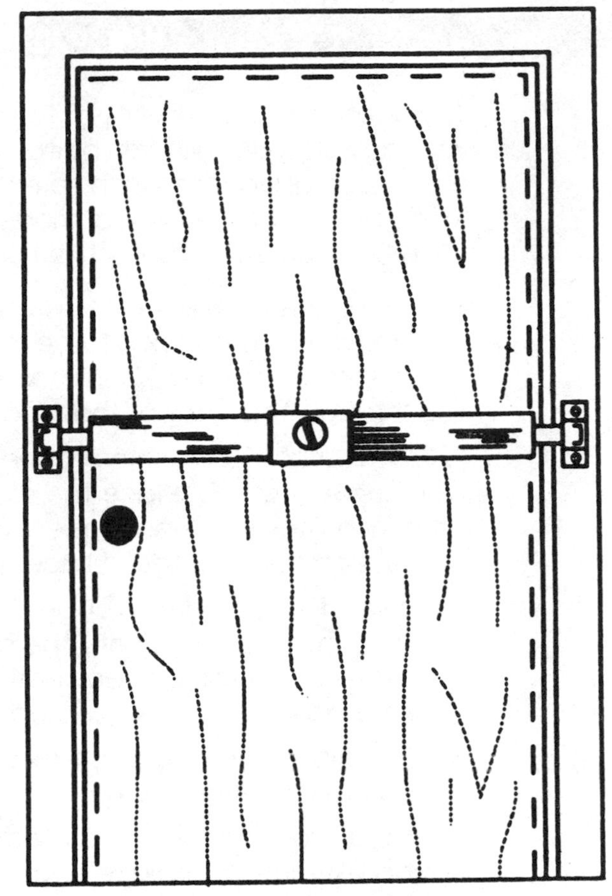

Figure 3–9. A "police lock" is a surface-mounted multiple-bolt lock that secures a door on the lock edge as well as the hinge edge. (Courtesy of Norden Lock Co., Inc.)

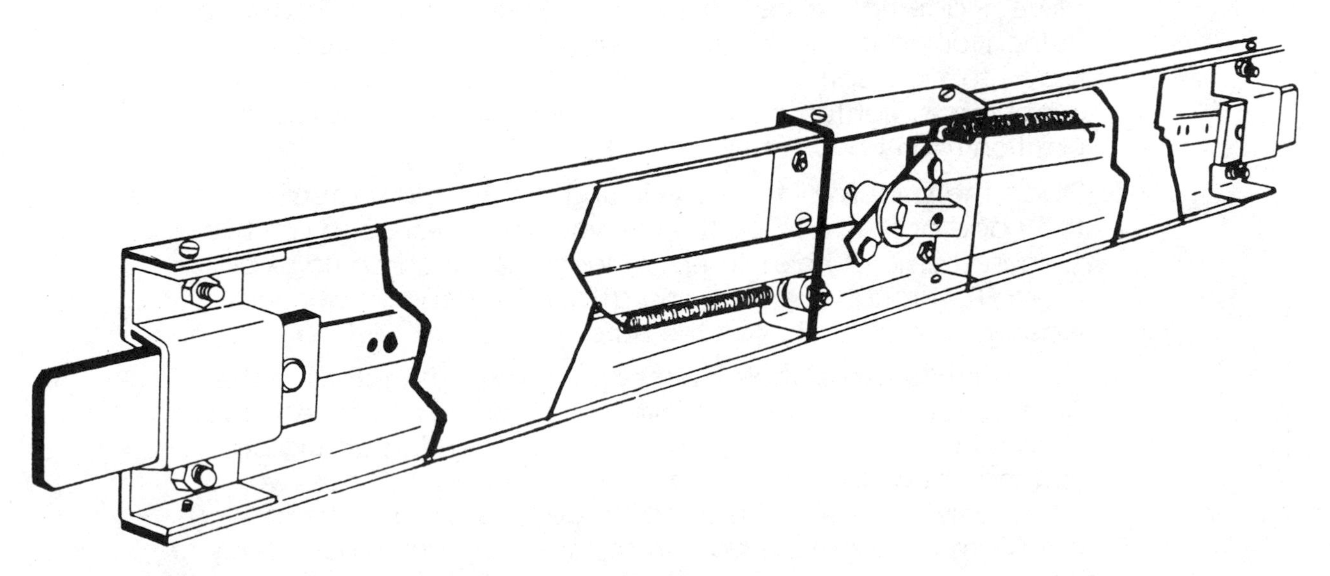

Figure 3–10. A cutaway view of a police lock. (Courtesy of Norden Lock Co., Inc.)

How to Install Police Locks

Although you may find its obtrusiveness aesthetically unappealing, you may decide that you need the strength of a police lock. You have several options when selecting a police lock. Here's the way to install one of the typical models. (See Figure 3–11 for the "items" (parts) specified in the instructions.)

1. Draw a level line across your door at the height where the bottom of the lock is to be installed. At the midpoint of that line, measure upward 2 1/4 inches. Drill a 1 1/2-inch-diameter hole through the door at that spot.
2. With a hacksaw, trim off one section of both cylinder retaining screws (item F). Place the screws through the interior backing plate (item I). Then place spacers (item E) onto the screws and attach them to the cylinder (item D).
3. Insert the cylinder through the drilled hole in the door and screw the backing plate onto the door with 1-inch pan head screws (item G). Break off two sections (about 1/2 inch) of the tailpiece so that 1 inch of the tailpiece remains.
4. Using the backing plate as a template, drill 5/16-inch-diameter holes at the four corners of the plate.
5. Hold the exterior lock protecting plate (item A) long edge downward, and place two 1/4-inch × 2 1/4-inch carriage bolts (item B) in the upper right and lower left holes. Place two 1/4-inch × 3-inch carriage bolts (item C) in the upper left and lower right holes. Place the plate's attached carriage bolts into the holes drilled through the door and into the interior backing plate. Place flat washers onto the two 3-inch bolts and attach 1/4-inch lock nuts (item L) to all four bolts. Before tightening, place the two springs (item J) under the flat washers on the 3-inch bolts. Tighten the nuts just enough to keep the lock in position (don't overtighten).
6. Place the bottom of the two locking arm brackets (item EE) at the edge of the door, on the line you drew. Mark and drill two 3/8-inch-diameter holes. Insert 5/16-inch × 2 1/2-inch carriage bolts (item FF) through the door and attach the bracket with lock washers and nuts. Tighten the nuts.
7. Position the preassembled connecting arms (item P) and the pivot arm (item O) in place over the tailpiece. (Note: After you determine what length you need the connecting arm to be, you may have to cut it. But don't cut the end that has two holes; only cut the end that has three holes.) Place the locking bar (item AA) onto the bolt and spacer. This unit is now ready to install.

8. Place the slotted-groove tailpiece section (item M) onto the tailpiece, making sure that the groove is positioned to face the top. Place the nylon top hat spacer (item N) over the tailpiece with the flat side resting on the cylinder retaining screws. Insert the pivot arm section into the top hat spacer, making sure that the small hole is facing the top. (Note: The arms are in the right position if the top one points left and the bottom one points right.) The locking arms must be placed into their brackets while attached to the rest of the section. (This entire section can be lifted far enough from the backing plate to get it into position.)
9. You should now be able to move the locking arms by hand. Mark the place where the strike plate (item II) is to be cut into the jamb. After making the cut in the jamb, fasten the strike plate with the # 12 × 1 inch flat head sheet metal screws (item JJ).
10. Hold the two outer covers (item U) so that the screw holes line up with the holes in the locking arm supporting brackets (item EE). Mark the other end so that 3/8 inch of the outer cover rests over the interior backing plate (item I). (Make sure the covers don't cover the screw holes in the interior backing plate.) Secure the four 3/8-inch sheet metal screws (item Y) in the locking arm supporting brackets.
11. Position the bars into the locking position. If the upper bar, which should now be in its lowest position, doesn't rest against the 3-inch bolt, place a 3/4-inch × 3/8-inch spacer (unlabeled item) over the bolt and secure it with a 1/4-inch nut so that the arm now rests against it.
12. Spray the main mechanism with WD-40 or a similar type of lubricant.
13. Place the nylon spacer (item T) onto the pivot arm shaft. Then place the center cover (item V) over the shaft with the narrow part to the top. (If you look at the center cover, you will see that the hole is off-center; the shorter part is on top.) Secure the four 3/8-inch sheet metal screws (item Y) until the cover is attached.
14. Place the thumb-turn (item W) on the shaft and secure it with the oval head screw (item X).

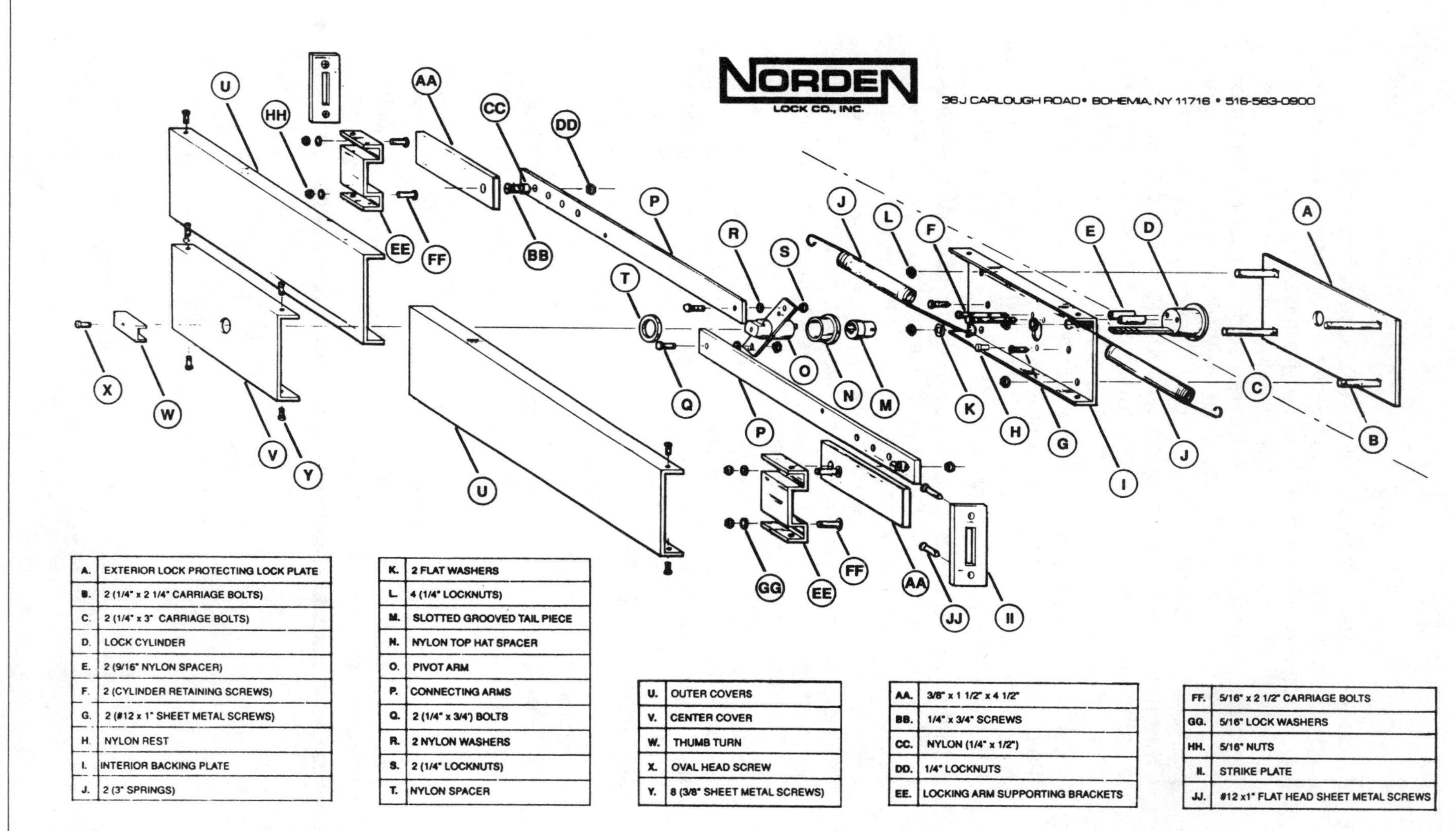

A.	EXTERIOR LOCK PROTECTING LOCK PLATE
B.	2 (1/4" x 2 1/4" CARRIAGE BOLTS)
C.	2 (1/4" x 3" CARRIAGE BOLTS)
D.	LOCK CYLINDER
E.	2 (9/16" NYLON SPACER)
F.	2 (CYLINDER RETAINING SCREWS)
G.	2 (#12 x 1" SHEET METAL SCREWS)
H.	NYLON REST
I.	INTERIOR BACKING PLATE
J.	2 (3" SPRINGS)

K.	2 FLAT WASHERS
L.	4 (1/4" LOCKNUTS)
M.	SLOTTED GROOVED TAIL PIECE
N.	NYLON TOP HAT SPACER
O.	PIVOT ARM
P.	CONNECTING ARMS
Q.	2 (1/4" x 3/4") BOLTS
R.	2 NYLON WASHERS
S.	2 (1/4" LOCKNUTS)
T.	NYLON SPACER

U.	OUTER COVERS
V.	CENTER COVER
W.	THUMB TURN
X.	OVAL HEAD SCREW
Y.	8 (3/8" SHEET METAL SCREWS)

AA.	3/8" x 1 1/2" x 4 1/2"
BB.	1/4" x 3/4" SCREWS
CC.	NYLON (1/4" x 1/2")
DD.	1/4" LOCKNUTS
EE.	LOCKING ARM SUPPORTING BRACKETS

FF.	5/16" x 2 1/2" CARRIAGE BOLTS
GG.	5/16" LOCK WASHERS
HH.	5/16" NUTS
II.	STRIKE PLATE
JJ.	#12 x1" FLAT HEAD SHEET METAL SCREWS

Figure 3–11. A typical police lock has many parts that can't be seen when the lock is installed. (Courtesy of Norden Lock Co., Inc.)

LOCK CYLINDERS

Regardless of the type of lock you choose, you'll need to decide whether you want a single- or double-cylinder model. A single-cylinder lock is one that uses a key on one side of the door and a thumb-turn on the other. A double-cylinder lock requires that a key be used on both sides of the door.

Security professionals disagree about which type of lock cylinder is best. Some point out that double-cylinder locks force a burglar who climbed in through a window to leave through a window—which lessens the amount that can be carried out. For that strategy to work, however, you would need to use double-cylinder locks on every door in your home, and you would have to install all of them in a way that would make removal difficult.

I don't recommend that strategy: it would present a serious problem if you needed to get out quickly—such as during a fire—and couldn't find the key. Your door locks should be for keeping unwanted people outside, not for trapping yourself inside.

Some lock makers, like Abloy High Security Locks and Kwikset, offer double-cylinder locks with safeguards. These devices allow the locks to work like single-cylinder models while people are inside a home. Never use double-cylinder locks that lack this safety feature for all the exterior doors of your home.

High-Security Cylinders

Any lock can be made stronger by replacing its standard cylinder with a high-security cylinder (see Figure 3–12). A high-security cylinder provides special protection against lockpicking, drilling, and other entry techniques that sophisticated burglars use. Although any lock that requires a key can eventually be picked-open, locks with high-security cylinders are almost never defeated by lockpickers.

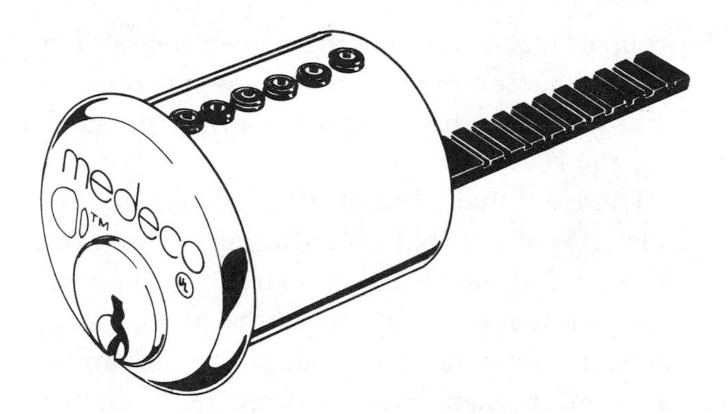

Figure 3–12. Although from the outside a high-security cylinder doesn't look much different from a standard cylinder, it is designed to thwart drilling, lockpicking, and other sophisticated entry techniques. (Courtesy of Medeco Security Locks)

Ordinary lock picks are seldom successful on a high-security cylinder. Even with custom-made tools, it could take a professional burglar (or a locksmith) several hours to pick-open a typical high-security cylinder. Few home burglars are willing to spend that much time at one place.

An important advantage in using a high-security cylinder is that you'll have maximum control over who may get a copy of your key. The easier a key can be duplicated, the easier a lock can be compromised. If you're like most people, you've entrusted your house keys at one time or another to friends, neighbors, a home-repair contractor who needed access at a time when you could not be home, a parking-lot attendant (if you keep all your keys on the same key ring), and others. With standard lock cylinders, you never know when someone might make copies of your keys at the nearest hardware store.

Keys for high-security cylinders are hard for unauthorized persons to duplicate, because they can be copied only on special key machines. Manufacturers of high-security cylinders often use patented keys to further ensure key control. Patented keys can be duplicated only by a select group of locksmiths who have been approved by the manufacturers and who will require proof of lock ownership from anyone wishing to duplicate the keys.

There are many brands of high-security cylinders; most are sold only through locksmiths. Because each line involves a large initial investment for merchandise, special machines, special tools, training, and the like, few locksmiths carry more than two or three high-security product lines. Don't be surprised if the locksmith you approach tries to explain in great detail why the one line he carries is unique and superior to all others.

The major high-security cylinders—those made by Abloy, Assa, Medeco, and Schlage Primus, to name a few—have unique, patented features. Your only concern should be whether the cylinder is listed by Underwriters Laboratories (UL), an independent testing agency. Any UL-listed model can easily meet or exceed the needs of most homeowners. A UL listing means that a sample model has withstood rigorous expert attack tests.

You can buy a high-security cylinder for under $100. Make sure the one you select fits the lock it is intended for. Find out how much you'll have to pay for duplicate keys. Depending on the brand, you can expect to pay between $3 and $10 per key.

Making a Standard Cylinder More Secure

For about $10, you can make a standard cylinder "almost high security." Just have a locksmith re-key the cylinder using two mushroom or spool pins, and the cylinder will be more pick-resistant. Ask the locksmith to make your new keys on high-security key bow (rhymes with "toe") blanks. This type of key blank doesn't show the information that most key cutters need to duplicate keys. Only highly experienced locksmiths know how to duplicate a key made on a high-security key bow blank. The fewer people who know how to duplicate your key, the harder it will be for someone to quickly have a duplicate made.

MINIMIZING YOUR KEYS

Four thousand years after door keys were invented, we still rely on them for security. Unlike the large wooden keys the early Egyptians proudly carried on their shoulders, today's small metal keys are often more a burden than a source of pride.

They tear holes in pockets, they're inconvenient to carry around, and they're easy for children and adults to lose. If you're like most people, you have to fumble through eight or more keys to get into your home. There are simple ways to reduce your need for keys—without reducing your security.

One way to make your key ring smaller is to have all or most of your locks work with the same key. Some people have several door locks that use different keys because they think burglars have more trouble picking-open a variety of locks. Locks that are keyed alike are no easier to defeat than those that are keyed differently. They still have to be picked-open one at a time.

If you want the convenience of using one key for your doors, some of your locks may need to be rekeyed (so that all of them have the same tumbler pattern). Locksmiths charge about $10 per lock for rekeying.

Sometimes, it isn't practical to make two locks fit the same key. If the key to one can't slide into the keyway of another, for example, then those two locks can't share a key. Locks come in hundreds of different keyway shapes and sizes, and each can accept only certain keys.

You can avoid the hassle and extra costs of re-keying if you consider key compatibilities when you're buying new locks. Many locks sold at hardware and department stores have keying numbers (or "tumbler pattern numbers") on their packages. You may be able to find several locks with the same number, which means they're keyed alike. (Because there are so many possible keying numbers, when you see the same numbers on more than one package you can assume those locks also have the same keyway.)

If you use padlocks around your home, you may want to get models that will work with your door key. You can find them at locksmith shops. Bring your door key with you to make sure the padlock has the right keyway, and have it re-keyed to match your door lock.

The exact cost will depend on the size, type, and brand of lock you choose, but expect to pay at least $20 for any padlock that uses your door key. If you don't want to spend that much, you can buy a good combination padlock for much less. (If the body and shackle of a combination padlock are strong, it can provide as much protection as a key-operated padlock.)

A combination padlock isn't the only keyless lock you can use in your home. Several companies make pushbutton locks for exterior doors (see Figure 3–13). These locks are especially useful for young children or anyone who has a hard time keeping track of keys. Make sure any keyless lock you install on an exterior door is weather-resistant and has a rigid bolt that projects from the edge of the door. Stay away from exterior door locks with spring-loaded bolts only. They're easy to defeat. (If the bolt is beveled, you can assume it's spring-loaded.)

How to Install Pushbutton Door Locks

Specific installation methods vary among brands and models, but most pushbutton door locks that are designed to be mortise-mounted can be installed by following these four steps:

1. When you decide the height at which you want the lock installed—the usual height is about four feet from the bottom of the door—position the lock's template at the desired height and use it to mark the drill points on the door.
2. Bore the openings for the lock and latch (see Figure 3–14).
3. Mortise the area for the bolt face, insert the bolt housing (see Figure 3–15), and screw the bolt housing into place.
4. Insert the lock and tighten the screw connecting the bolt housing to the lock (see Figure 3–16). Using the screws provided, attach the cover plate on the inside of the door.

If a door has holes that were drilled for a previous lock, you can use escutcheon plates to cover any visible unneeded holes. These plates come in a wide variety of finishes and styles. (See Chapter 2 for information about other products that can be installed along with locks.)

Electronic Locks

The newest type of combination locks being used on homes are key-optional electronic models. Most of them are reliable and offer advantages over a mechanical lock, such as built-in alarms or linkage to an alarm system. They allow you to change your access code quickly, at any time, without going to a locksmith. These models sell for between $100 and $300.

The main problem with electronic locks is that you have to be sure to change their batteries periodically. Some models, like InteLock's model 3000, give a low-battery signal about 3

Figure 3–13. A rugged weather-resistant keyless lock with a nonbeveled bolt at least 1 inch long can offer as much protection as a tubular deadbolt lock. (Courtesy of Preso-Matic Lock Company)

Figure 3–14. When boring a large cavity for a mortise-mounted lock, you have to use a large bit to drill in several places. (Courtesy of Preso-Matic Lock Company)

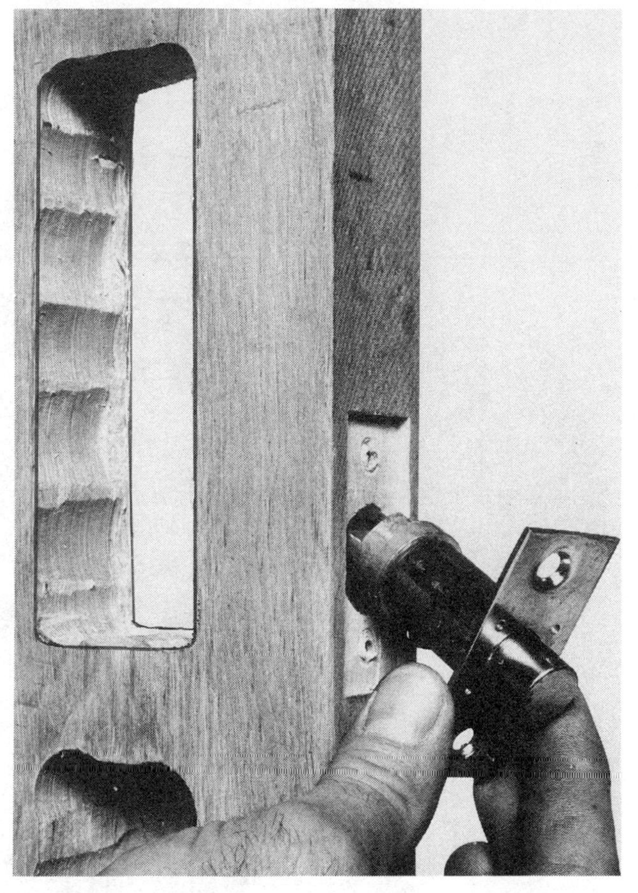

Figure 3–15. As is the case with most other door locks, the bolt assembly of a Preso-Matic brand keyless lock needs to be installed in a door before other parts are installed. (Courtesy of Preso-Matic Lock Company)

months before the batteries need to be replaced (see Figure 3–17).

The InteLock 3000, which looks like an ordinary deadbolt and key-in-knob set of locks, can be retrofitted into existing lock holes. It can be unlocked by first twisting the knob and then twisting the deadbolt. As the knob is turned left and right, digits appear in the electronic display on the lock's escutcheon. The user twists the knob until the desired personal code has been entered, and then turns the cylinder guard on the deadbolt, to open the lock.

You can program a temporary code any time to allow, say, a baby-sitter, to get in without knowing the master code. If someone tries to guess the code by rotating the key-in-knob too much, an alarm will sound.

Figure 3–16. Before attaching the cover plate to a Preso-Matic brand keyless lock, make sure the bolt extends and retracts properly. (Courtesy of Preso-Matic Lock Company)

Figure 3–17. Although it looks like a standard deadbolt and key-in-knob, the InteLock 3000 can be unlocked by twisting the knob left and right to insert a code. (Courtesy of InteLock Corporation)

RECOMMENDED RESOURCES

(See HM Sourcelist for more information.)

Abloy High Security Locks—High-security cylinders and locks

Assa High Security Locks—High-security cylinders and locks

DOM Security Locks—High-security cylinders

Fichet-Brauner, USA—High-security cylinders and locks

InteLock Corporation—Electronic door locks

Kwikset Corporation—Mechanical door locks, padlocks

Master Lock Company—Mechanical door locks, padlocks

Medeco Security Locks—High-security cylinders and locks

Mul-T-Lock Corporation—High-security cylinders and locks

The New England Lock & Hardware Company—Mechanical cylinders and door locks

Norden Lock Company, Inc.—Mechanical door locks

Preso-Matic Lock Company—Pushbutton door locks

Schlage Lock Company—High-security cylinders and locks, mechanical door locks

Simplex Access Controls—Pushbutton door locks

VSI Donner—Padlocks

Yale Locks & Hardware—Mechanical cylinders and door locks

4

The Safest Safes

Important documents, keepsakes, and other hard-to-replace valuables need special protection from theft and fire. A low-cost safe may provide the protection you need for them. Not all safe designs are alike. Whether a safe costs $100 or $10,000, using the wrong one can be a costly mistake.

Safes vary greatly in quality, price, and usefulness. To find one that meets your needs at the best price, ask yourself these four questions:

1. What kind of protection do I need?
2. How much protection do I need?
3. Where will I put the safe?
4. What special features do I want?

TYPES OF PROTECTION

Most safes provide strong resistance to fire or burglary, but not to both. The construction that makes one type of safe fire-resistant—thin metal walls with insulation sandwiched between—also makes it easy to break into. The construction that makes another type hard to break into—thick, heavy, steel walls—makes it conduct heat easily.

A so-called "burglary/fire" safe is two safes in one: a small burglary safe welded into a fire safe. It's generally less expensive to buy two safes instead of one burglary/fire model.

The kind of safe you need depends on what you want to protect. For your tax records and other personal documents, for example, you probably don't need a burglary-resistant model because a burglar isn't likely to want your papers.

DEGREES OF PROTECTION

It's important to understand that there is no universally accepted standard for terms like "fire-resistant" and "burglary-resistant." "Resistant" is a broad, undefined description. Each safe manufacturer can choose when and how it will use these terms, regardless of how little resistance a particular safe provides. Many safes that are advertised as "burglary-resistant" can be easily pried open with a screwdriver. Be sure that the safe you plan to buy meets standards set by an independent rating agency.

Two well-known independent agencies are Underwriters Laboratories (UL) and Japanese Industrial Standards (JIS). Each agency offers a wide range of ratings indicating how much protection a safe provides. If a safe model has been rated, it will have a rating label. Although UL and JIS have many equivalent standards for safes, the UL label is more commonly found on models sold in the United States.

For a fire safe to earn a UL (or JIS) rating, it must meet strict construction guidelines and a sample model must pass several rigorous tests that simulate what a safe might go through in a fire. The tests include: an explosion hazard test, a fire endurance test, a fire and impact test, and a humidity test.

A fire safe's rating is based on the maximum temperature that may be reached inside its insulated area during a certain period of time. UL ratings include: 350-1, 350-2, 150-1, 150-2, 125-1, and so on. The number to the left of the hyphen represents the maximum temperature that can be expected inside the safe during a typical house fire. The number to the right of the hyphen represents the maximum amount of time you can expect the safe to protect its contents. In other words, a 350-1 rating indicates that the inside temperature of the safe shouldn't exceed 350 degrees Fahrenheit during the first hour of a typical house fire. A 350-2 rating offers that level of protection for 2 hours.

The rating you need depends primarily on what you want to store in the safe. All 350-class safes are useful for storing documents, because paper doesn't char until it's exposed to 405 degrees Fahrenheit. But you'll need a 125-class safe (called a "media safe") to protect phonograph records, audio tapes, compact disks, and computer disks. They would be ruined if they were exposed to any higher temperatures.

There are also UL and JIS ratings for burglary safes. To earn one of these ratings, a burglary safe must meet strict construction guidelines regarding weight, materials, thickness of walls, type of lock, and other structural specifications, and a sample model must have withstood an attack by a safe expert using certain tools for a specified period of time. Various burglary ratings are based on what types of tools were used in the test and how long the safe withstood the attack.

The ratings include: TL-15, TL-30, TRTL-30, and TRTL-60. "TL" means the safe resisted attack by an expert using common mechanical, electrical, and hand tools. "TRTL" means the safe withstood all those tools plus cutting torches. The number that follows the hyphen tells how many minutes the safe withstood the attack. Depending on its type and size, a TL-15 safe can cost between $700 and $7,000. Because of their price range, UL-rated safes are used more often in businesses than in homes.

Many burglary safes used in homes are classified as having one of the Broad Form & Mercantile Safe Insurance Classifications: B or C. (Both classifications offer less burglary resistance than a TL-15 safe.) B-rated models have a steel or iron door less than 1 inch thick, and steel or iron walls less than 1/2 inch thick. C-rated safes have a steel door at least 1 inch thick, and steel walls at least 1/2 inch thick.

How to Install a Wall Safe

Most wall safes can be installed in a home by doing the following:

1. Locate the wall studs.
2. Mark the wall, using a template as a guide.
3. Cut out the drywall panel, using a small saw.
4. Remove the cut-out portion of drywall.
5. Insert the safe into position between the studs.
6. Place screws through the holes of the safe and screw them into the studs.

WHERE AND HOW TO INSTALL SAFES

Most safes for homes are designed to be installed in a wall, on a floor, or in a floor. They are called, respectively, wall safes, floor safes, and in-floor safes.

Wall Safes

Wall safes provide a convenient and inconspicuous place to store valuables. Although some models are fire-resistant, none should be considered burglary-resistant if it is made for residential use. Because wall safes have to be light enough to rest in a drywall cutout, an entire safe can be easily yanked out. If the safe is hard to force open, a burglar can just carry it away and work on it later, at leisure.

Even if you place a painting or mirror over it, your wall safe won't be safe from burglars. They almost always look behind objects hanging on a wall. A wall safe should be used only for storing lightweight items that a burglar wouldn't want. To prevent your walls from being broken up by burglars, you might want to write your safe's combination on the face of the safe—or leave the safe unlocked.

Floor Safes

If you need more burglary protection than a wall safe offers, or if you want to be able to remove your safe easily and take it with you when you move, you might consider getting a floor safe. (See Figure 4–1.)

A floor safe often has wheels and can be easily moved around. For security purposes, however, after the safe has been moved to its permanent location, the wheels should be removed. Some manufacturers recommend bolting to the floor any floor safe that weighs less than 750 pounds.

It's a good idea to put a floor safe in the corner of a room and anchor it to the floor and the two walls. Don't think a safe is secure just because it's heavy. Burglars have been known to use hydraulic hand trucks to take away safes weighing over 1,000 pounds.

How to Install an In-Floor Safe

In-floor safes are usually installed near walls to make it hard for burglars to use tools on them. Although procedures differ among manufacturers, most in-floor safes can be installed in an existing concrete floor in the following way:

1. Remove the door from the safe, and tape the safe's dust cover over the opening.
2. At the location where you plan to install the safe, draw the shape of the body of the safe, allowing 4 inches extra width on each side. (For a square-body safe, for example, the drawing should be square, regardless of the shape of the safe door.)
3. Use a jack hammer or large hammer drill to cut along your marking.
4. Remove the broken concrete, and use a shovel to make the hole about 4 inches deeper than the height of the safe.
5. Line the hole with plastic sheeting or a waterproof sealant (to resist moisture buildup in the safe).
6. Pour a 2-inch layer of concrete in the hole to give the safe a level base to sit on.
7. Place the safe in the center of the hole and shim it to the desired height.
8. Fill the hole with concrete all around the safe, and use a trowel to level the concrete with the floor. Allow 48 hours for it to dry.
9. After the concrete has dried, trim away the plastic and remove any excess concrete.

Figure 4–1. A small floor safe can be easily moved from one place to another. (Courtesy of Gardall Safe Corporation)

In-Floor Safes

If you need a safe that provides a lot of fire and burglary resistance and you want it to be permanently installed, your best choice might be an in-floor (or "in-the-floor") safe. They come in different sizes (see Figure 4–2) and can be installed in new or existing concrete or wood floors (see Figure 4–3).

Some models have UL burglary ratings, but they don't meet UL construction requirements for fire ratings. However, because heat rises during a fire and in-floor safes are installed below floor level, little heat builds up in a floor safe during a typical house fire (unless the safe is installed on an upper floor).

In-floor safes aren't problem-free. They collect dirt and moisture and need to be serviced

Figure 4–2. In-floor safes come in various sizes. (Courtesy of Gardall Safe Corporation)

regularly. They're also inconvenient: you have to sit or crouch on the floor and bend over to use them. Anyone with a bad back may have a hard time using an in-floor safe.

SPECIAL FEATURES TO LOOK FOR

An important feature to consider when buying a safe is the type of lock it uses. The traditional combination lock with a dial ("36 to the left, 93 to the right . . .") is the most common type, but many people prefer the convenience of a key-operated lock.

Some safes use pushbutton mechanical or electronic locks. Both types are easy to operate and allow quick access to the safe. The electronic models are not popular for home use because they generally cost more and they run on batteries. People don't like having to remember to replace batteries.

When choosing a burglary safe, consider whether it has relocking devices ("relockers") and how many it has. Relockers are spring-loaded supplemental locks that are automatically triggered into the locked position when a burglar attempts to force open the safe's primary lock. Once a relocker has been activated, even a locksmith will have a hard time getting the safe open. Another feature found in good burglary safes is hardplate—special material that hinders thieves from drilling the safe open. At the very least, a safe should have hardplate between the safe's lock and dial. The more hardplate, the better.

SPECIALTY SAFES

If you need a safe for protecting a specific type of item, you may be able to use a specialty safe. A gun safe, for example, is shaped especially for holding rifles. Other specialty safes include media chests—for holding computer disks—and file cabinet safes. (See Figures 4–4 and 4–5.)

GETTING A GOOD BUY

You'll find the lowest priced safes at department stores and hardware stores. Most of them will be low-rated or nonrated fire safes with few security features. Remember: you don't get what you don't pay for. High-quality safes cost more to make than low-quality safes do.

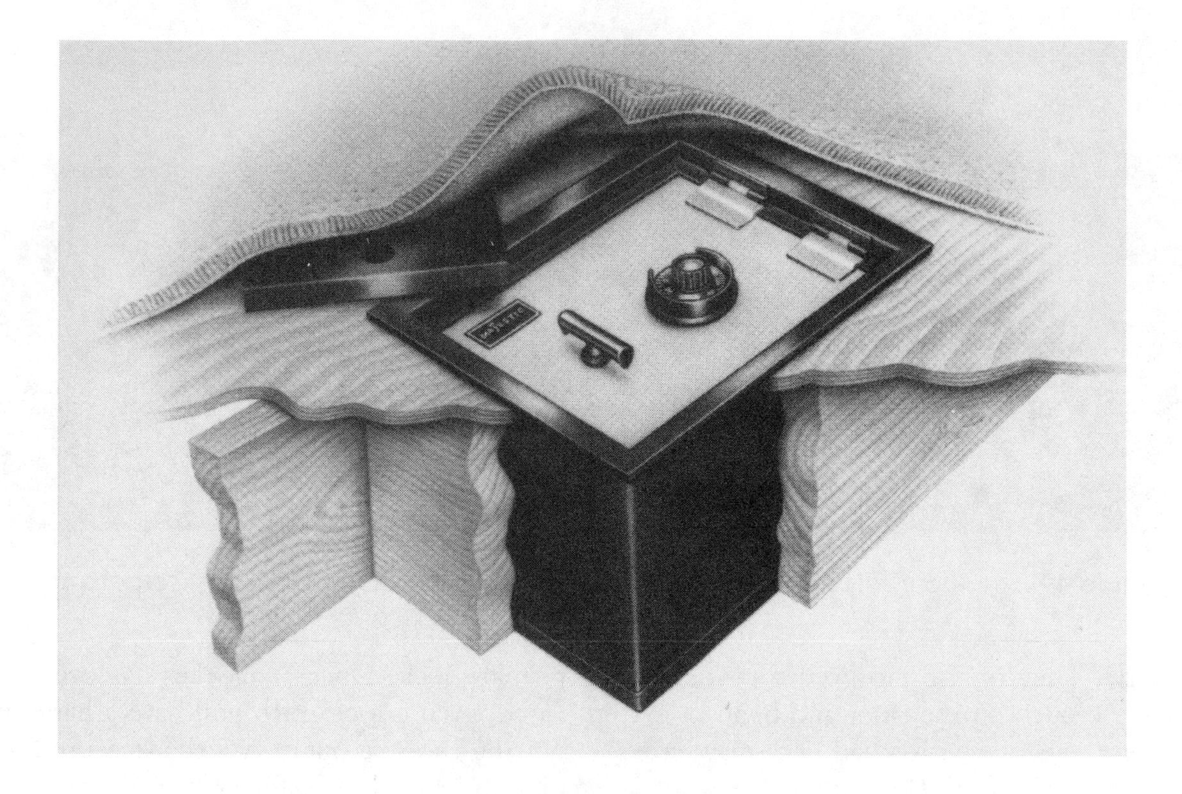

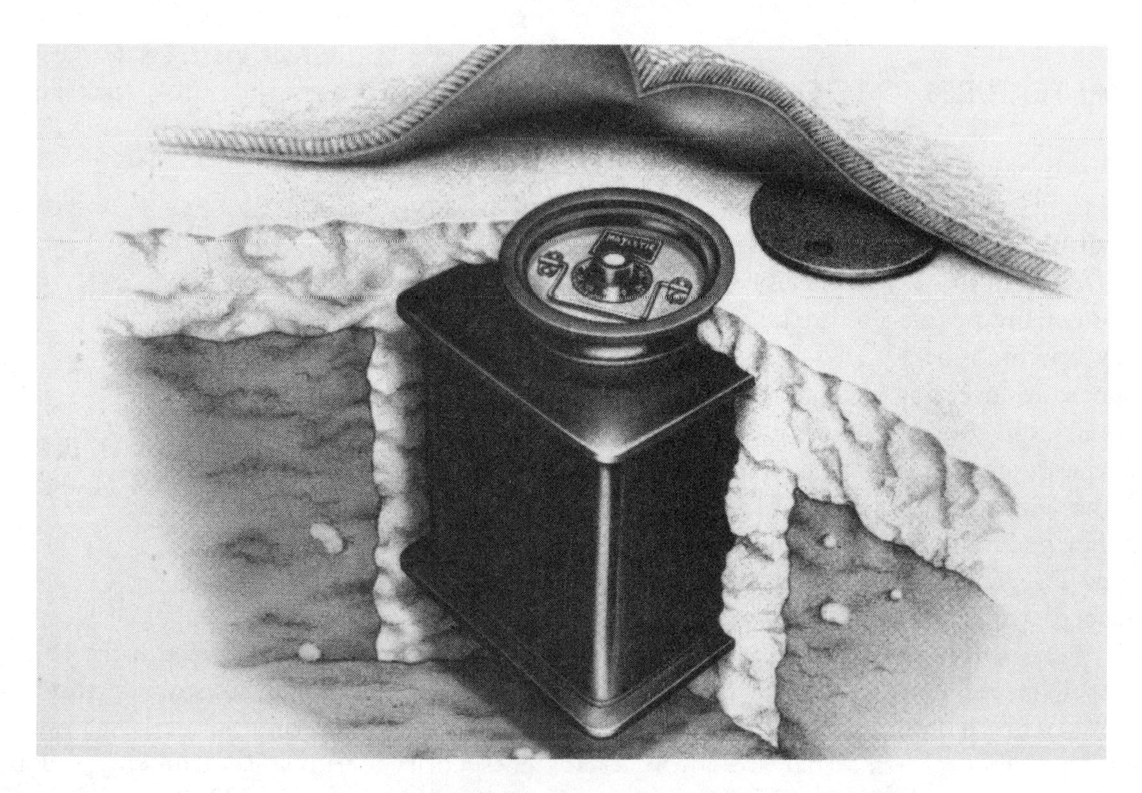

Figure 4–3. In-floor safes can be installed in wood or concrete floors. (Courtesy of Majestic Company)

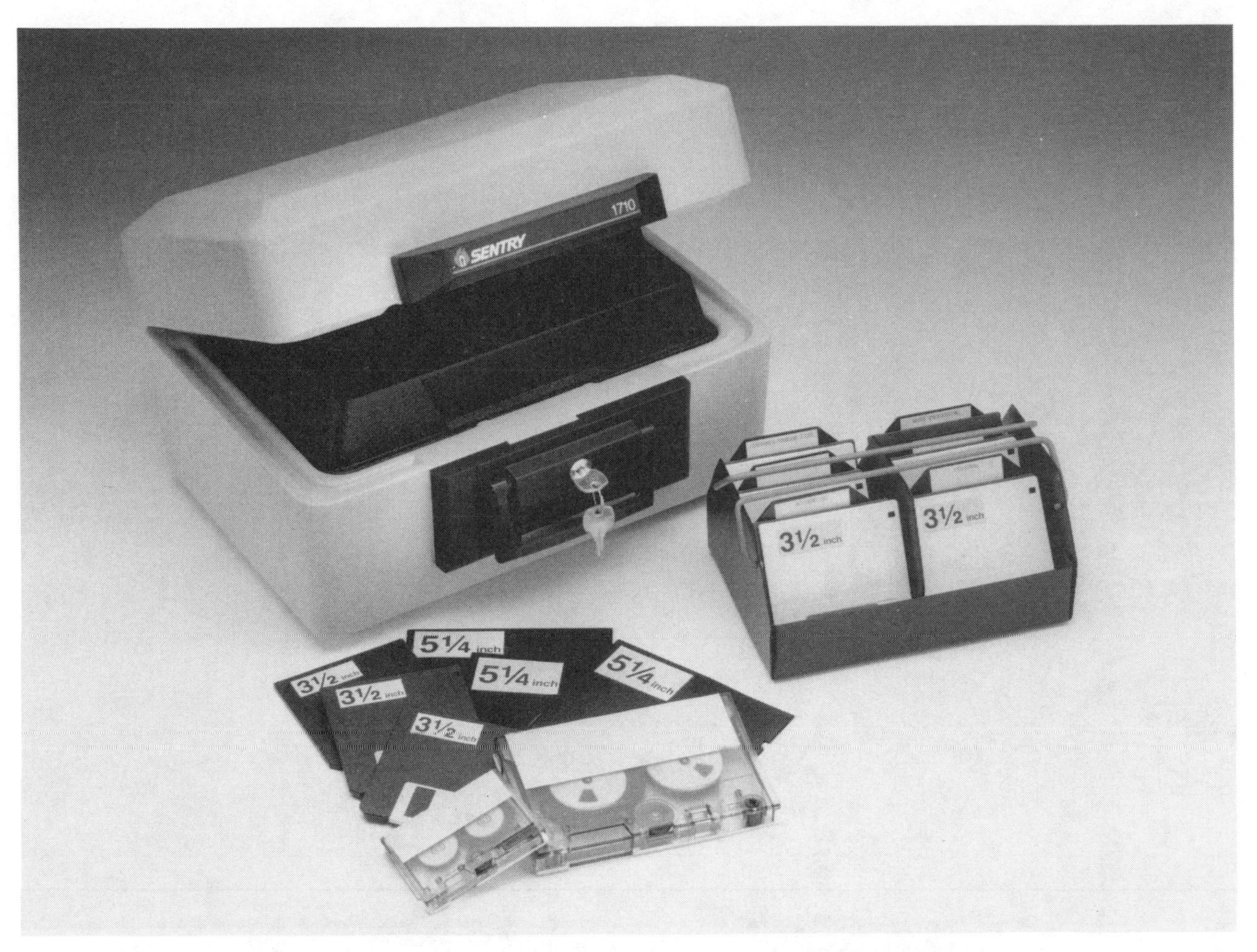

Figure 4–4. A media safe is designed to protect computer disks from fire. (Courtesy of Sentry Group)

At a typical locksmith shop, you'll find only a few safes. They will usually be more expensive, but of higher quality, than those sold in department stores. For the widest selection and most competitive prices, go to a shop that specializes in selling and servicing safes. Look for one listed under "Safes & Vaults" in the Yellow Pages of your local telephone directory.

The shop may sell used safes at bargain prices. Be careful about how "used" a safe has been. It isn't a good idea to buy a fire safe that has been through a fire or a burglary safe that has been drilled open. Even if they've been repaired, these safes can never offer as much protection as they once did.

When comparing prices among safes, ask about delivery and installation costs. Both are usually extra.

ALTERNATIVES TO A SAFE

It's important to remember that all home safes are a compromise between security and convenience. No ordinary home is built to withstand the kind of physical attacks that banks and similar buildings can withstand. If you don't have to store valuable or hard-to-replace items at home, you may not need a home safe.

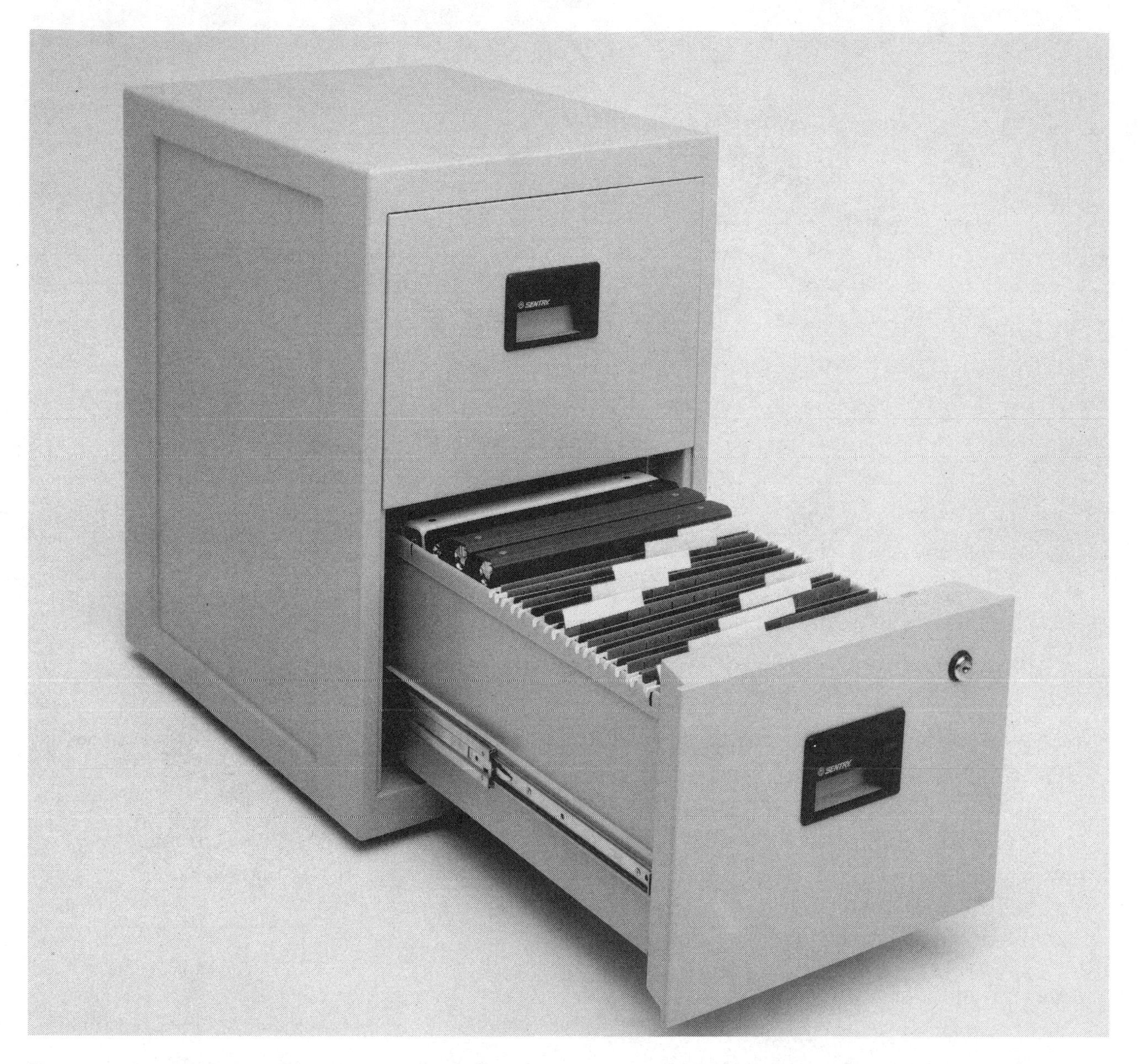

Figure 4–5. A file cabinet safe can be a convenient way to protect papers from fire. (Courtesy of Sentry Group)

It may be better to rent a safe deposit box at a bank. The smallest boxes are usually about 3 inches × 5 inches × 17 inches, and may be available for less than $20 per year, depending on where you live. The problem with bank safe deposit boxes is that you can't get to your items when the bank is closed.

If you need 24-hour access to your valuables, you might want to store them in a private vault. Private vaults come in a wide variety of sizes, and most are as secure as bank vaults. Private vaults usually cost more to rent. Expect to pay at least $50 for the smallest container at a private vault. For a list of private vaults near you, contact the National Association of Private Security Vaults. The organization's strict certification standards are tougher than the standards of many states.

RECOMMENDED RESOURCES

(See HM Sourcelist for more information.)

Buddy Products—Wall safes

Cannon Safe—Fire and burglary safes

Fort Knox—Fire safes

Gardall Safe Corporation—Fire and burglary safes

Majestic Company—Fire and burglary safes

National Association of Private Security Vaults—Trade association

Sentry Group—Fire safes, insulated file cabinets, media chests

5

Electronic Security for Today and the Future

Nowhere have recent advances in electronic and computer technology been more apparent than with security systems. Many types of systems that sell for under $1,000 today weren't available 10 years ago at any price, and some of today's lowest priced systems are more effective and more reliable than ever.

To get your money's worth, however, you have to know what to look for. This chapter reviews a wide range of electronic security systems and devices. I'll explain why some of them can be useful, and why many others can be costly nuisances. I'll also show you the basic installation procedure used for many types of wireless alarms.

BURGLAR ALARMS

More than 600 inmates of an Ohio prison were asked what single thing they would use to protect their homes from burglars. The most popular choice was a dog; the next was a burglar alarm. Other studies show that many police officers also believe a burglar alarm can make a home safer.

I favor installing burglar alarms, but they're not useful for everyone. To benefit from a burglar alarm, you and everyone in your home must learn how to operate it properly and must use it consistently. Everyone must remember to keep all windows and doors of the house closed. Many homeowners pay thousands of dollars for an alarm system only to discover that using it is too much trouble.

Contrary to popular belief, a burglar alarm doesn't stop or deter burglars. It only warns of their presence (if it's turned on during a break-in). Some burglar alarm sellers say that your having an alarm will make burglars think twice about trying to break into your home. Actually, it isn't your having the alarm that deters burglars; it's their belief that your home has an alarm that will stop them. Often, the only part of a burglar alarm that can be seen from outside is the window sticker. If you use alarm system window stickers, very few burglars will know whether you have or don't have a burglar alarm.

If you're like me and you have to have the real thing, you can either buy an alarm as a complete kit or get the components separately. The components are likely to include a control panel, a siren or bell, and various detection devices. In general, complete kits are less expensive than separate components, but by mixing and matching components you can create a more effective burglar alarm system.

Detection devices (or "sensors") are the eyes and ears of a burglar alarm system. They sense the presence of an intruder and relay the information to the system's control panel, which activates the siren or bell. Today, you have more detection devices to choose from than ever before, but if you use the wrong type the wrong way, you'll have a lot of false alarms.

Some detection devices respond to movement, some to sound, and others to body heat. The principle behind each type is similar: When an alarm system is turned on, the devices sense and monitor a "normal" condition; when someone enters a protected area, the devices sense a disturbance in the normal condition and trigger an alarm.

Most detection devices fall within two broad categories: perimeter and interior. Perimeter devices are designed to protect a door, window, or wall. They detect an intruder before entry into a room or building. The three most common perimeter devices are foil, magnetic switches, and audio discriminators.

Interior (or "space") devices detect an intruder upon entry into a room or building. The five most common interior devices are ultrasonics, microwaves, passive infrareds, quads, and dual techs.

Foil

You've probably seen foil on storefront windows. It's a thin, metallic, lead-based tape, usually 1/2 inch to 1 inch wide, that's applied in continuous runs to glass windows and doors. Sometimes, foil is used on walls. Like wire, foil acts as an electrical conductor to make a complete circuit in an alarm system. When the window (or wall or door) breaks, the fragile foil breaks, creating an incomplete circuit and triggering the alarm.

Usually, foil comes in long, adhesive-backed strips and is applied along the perimeter of a sheet of glass or drywall. Each end of a run must be connected to the alarm system with connector blocks and wire. Foil is popular in stores because it costs only a few cents per foot.

There are three major drawbacks to foil:

1. It can be tricky to install properly.
2. It breaks easily when a window is being washed.
3. Many people consider it unsightly.

Whether or not you like foil, foil alone is rarely enough to protect a home. Other detection devices should also be used.

Magnetic Switches

The most popular type of perimeter device is the magnetic switch (see Figure 5–1). It's used to protect doors and windows that open. Magnetic switches are reliable, inexpensive, and easy to install.

As its name implies, the device consists of two small parts: a magnet and a switch. Each part is housed in a matching plastic case. The switch contains two electrical contacts and a metal spring-loaded bar that moves across the contacts when magnetic force is applied. When magnetic force is removed, the bar lifts off one of the contacts, creating an open circuit and triggering an alarm condition.

Figure 5–1. A magnetic switch is a low-cost detection device for doors and windows that open. (Courtesy of X-10 (USA) Inc.)

In a typical installation, the magnet is mounted on a door or window, and the switch is aligned about 1/2 inch away on the frame. When an intruder pushes the door or window open, the magnet is moved out of alignment.

Some magnetic switches are rectangular, for surface mounting. Others are cylindrical, for recess mounting in a small hole. The recess-mounted types look nicer because they're less conspicuous, but they're a little harder to install.

One problem with some magnetic switches is that an intruder can defeat them by using a strong magnet outside a door or window to keep the contacts closed. Some models can be defeated by placing a wire across the terminal screws of the switch and jumping the contacts. Another problem is that, if a door is loose fitting, the switch and magnet can move far enough apart to cause false alarms.

"Wide gap" reed switches can be used to solve those problems. Because reed switches use a small reed instead of a metal bar, they're less vulnerable to being manipulated by external magnets. The wide gap feature allows a switch to work properly even if the switch and magnet move from 1 inch to 4 inches apart. Some magnetic switches come with protective plastic covers over their terminal screws. The covers thwart attempts at jumping. Most types of magnetic switches cost less than $4 each.

Audio Discriminators

Audio discriminators trigger alarms when they sense the sound of glass breaking. The devices are very effective and easy to install. According to a survey by *Security Dealer* magazine, over 50 percent of professional alarm installers favor audio discriminators over all other forms of glass-break-in protection.

By strategically placing audio discriminators in a protected area, you can protect several large windows at once. Some models can be mounted on a wall up to 50 feet away from the protected windows. Other models, equipped with an "omnidirectional pickup pattern," can monitor sounds from all directions and are designed to be mounted on a ceiling for maximum coverage.

A problem with many audio discriminators is that they confuse certain high pitched sounds—like keys jingling—with the sound of breaking glass, and produce false alarms. Better models require both the sound of breaking glass and shock vibrations to trigger their alarm. That feature greatly reduces false alarms.

Another problem with audio discriminators is that their alarm is triggered only if glass is broken. An intruder can bypass the device by cutting a hole through the glass or by forcing the window sash open. Audio discriminators work best when used in combination with magnetic switches.

The devices usually cost between $30 and about $75 each.

Ultrasonic Detectors

Ultrasonic detectors transmit high-frequency sound waves to sense movement within a protected area. The sound waves, usually at a frequency of over 30,000 cycles per second, are inaudible to humans but can be annoying to dogs. Some models consist of a transmitter that is separate from the receiver; others combine the two in one housing.

In either type, the sound waves are bounced off the walls, floor, and furniture in a room until the frequency is stabilized. Thereafter, the movement of an intruder will cause a change in the waves and trigger the alarm.

A drawback to ultrasonic detectors is that they don't work well in rooms with wall-to-wall carpeting and heavy draperies because these soft materials absorb sound.

Another drawback is that ultrasonic detectors do a poor job of sensing fast or slow movements and movements behind objects. An intruder can defeat a detector by moving slowly and hiding behind furniture. Ultrasonic detectors are prone to false alarms caused by noises like a ringing telephone or jingling keys. Although they were very popular a few years ago, ultrasonic detectors are a poor choice for most homes today. They can cost over $60; other types of interior devices cost less and are more effective.

Microwave Detectors

Microwave detectors work like ultrasonic detectors, but they send high-frequency radio waves instead of sound waves. Unlike ultrasonic waves, these microwaves can go through walls and be shaped to protect areas of various configurations. Microwave detectors are easy to conceal because they can be placed behind solid objects. They are not susceptible to loud noises or air movement, when properly adjusted.

The big drawback to microwave detectors is that their sensitivity makes them hard to adjust properly. Because the waves penetrate walls, a passing car can prompt a false alarm. Their alarms can also be triggered by fluorescent lights or radio transmissions. Microwave detectors are rarely useful for homes.

Passive Infrared Detectors

Passive infrared (PIR) detectors (Figure 5–2) became popular in the 1980s. Today, they are the most cost-effective type of interior device for homes. A PIR senses rapid changes in temperature within a protected area by monitoring infrared radiation (energy in the form of heat). A PIR uses less power, is smaller, and is more reliable than either an ultrasonic or a microwave detector.

The PIR is effective because all living things give off infrared energy. If an intruder enters a protected area, the device senses a rapid change

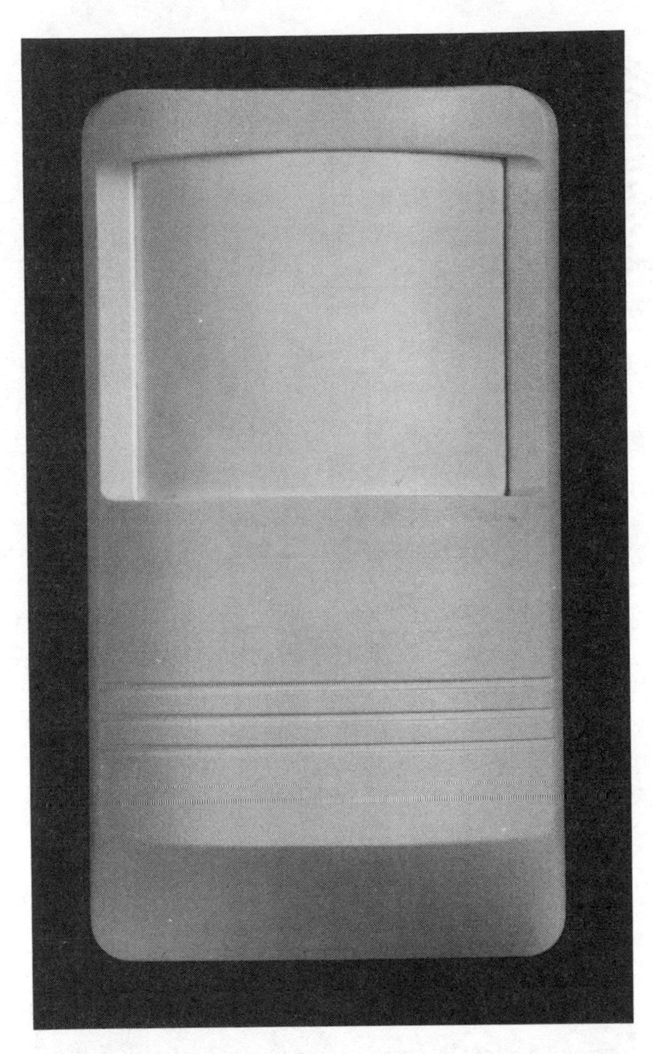

Figure 5–2. The PIR is the most cost-effective interior sensor for homes. (Courtesy of NAPCO Security Systems, Inc.)

in heat. When properly adjusted, the detector ignores all gradual fluctuations of temperature caused by sunlight, heating systems, and air conditioners.

A typical PIR can monitor an area measuring about 20 by 30 feet, or a narrow hallway about 50 feet long. It doesn't penetrate walls or other objects, so a PIR is easier to adjust than a microwave detector and it doesn't respond to radio waves, sharp sounds, or sudden vibrations.

The biggest drawback to PIRs is that they don't "see" an entire room. They have detection patterns made up of "fingers of protection." The spaces outside and between the "fingers" aren't protected by the PIR. How much of an area is monitored depends on the number, length, and direction of zones created by a PIR's lens, and on how the device is positioned (see Figure 5–3).

Many models have interchangeable lenses that offer a wide range of detection pattern choices. Some patterns, called "pet alleys," are several feet above the floor to allow pets to move about freely without triggering the alarm. Which detection pattern is best for you will depend on where and how your PIR is being used.

A useful feature of the latest PIRs is "signal processing" (also called "event verification"). This high-tech circuitry can reduce false alarms by distinguishing between large and small differences in infrared energy.

Quads

A quad PIR (or "quad," for short) consists of two dual-element sensors in one housing. Each sensor has its own processing circuitry, so the device is basically two PIRs in one. A quad reduces false alarms because, to trigger an alarm, both PIRs must simultaneously detect an intrusion. That feature prevents the alarm from activating in response to insects or mice. A mouse, for example, may be detected by the fingers of protection of one of the PIRs but would be too small to be detected by both at the same time.

Dual Techs

Detection devices that incorporate two different types of sensor technology into one housing are called dual technology devices (or "dual techs"). A dual tech triggers an alarm only when both technologies sense an intrusion. Dual techs are available for commercial and residential use, but because they can cost several

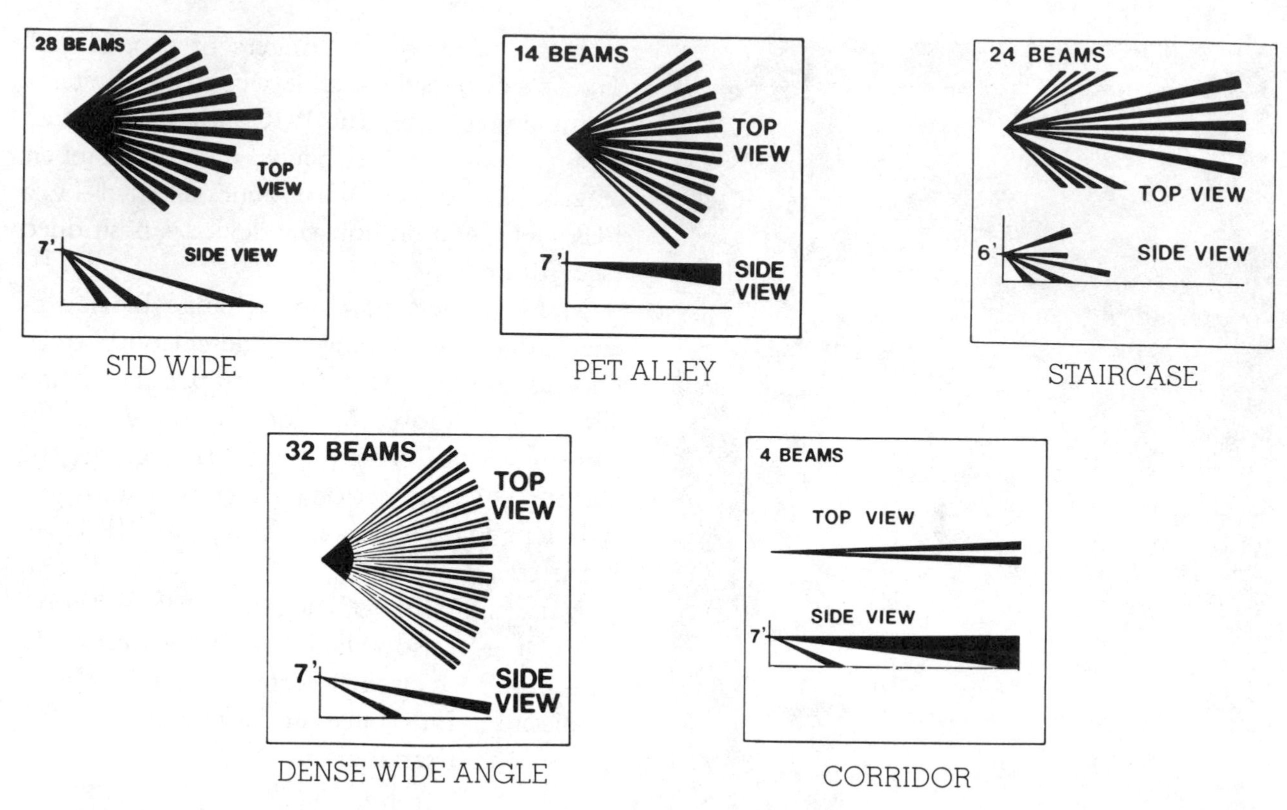

Figure 5–3. The lens of a PIR determines which spaces the device will be able to monitor. (Courtesy of NAPCO Security Systems, Inc.)

hundred dollars, dual techs are more often used by businesses. The most effective dual tech for homes is one that combines PIR and microwave technology.

For this type of dual tech to trigger an alarm, a condition must exist that simultaneously triggers both technologies. The presence of infrared energy alone, or movement alone, would not trigger an alarm. Movement outside a wall, which might ordinarily trigger a microwave, for example, won't trigger a dual tech because the PIR element wouldn't simultaneously sense infrared energy.

PROFESSIONAL VERSUS DO-IT-YOURSELF ALARMS

Until the past few years, there were sharp differences between professional and do-it-yourself home alarms. A professional system was more reliable and harder for burglars to defeat. Do-it-yourself models were cheaper and easier to install, but weren't very reliable. They often produced false alarms and failed to detect intruders.

Many of today's do-it-yourself systems, however, are not only simple to install, but are also very effective and reliable. If you know what to look for, you may find that a do-it-yourself model is a better buy for you than a comparable professionally installed model.

Few installers buy home alarms in large enough volume to compete with the prices of hardware, department, and electronic products chain stores. To make it harder for consumers to compare an alarm installer's prices with those found in chain stores, many installers sell lesser known (but more expensive) brands. That helps installers to promote a sales pitch claiming that although their systems cost more, they're better.

The "hardwired" alarm system has always been the alarm of choice among professional installers. This system requires wire to be run from its control panel to each of its sensors (and to its siren or bell). Hardwiring will give you maximum reliability and will allow you quickly to find and stop false alarms. However, running wires in a way that is both aesthetically pleasing and hard for burglars to defeat can be tricky and time-consuming. For this reason, few hardwired models are made for do-it-yourselfers.

The wireless alarm is easy to install and therefore is very popular among do-it-yourselfers (see Figure 5–4). It relies on radio waves—instead of wire—for communication between its control panel and sensors. Some wireless alarms can be installed with nothing more than a screwdriver. (With some "wireless" models, you still have to run wire from the siren to the control panel.)

Until recently, the biggest problem with do-it-yourself wireless alarms was that they weren't "supervised"; only expensive high-tech models included supervising circuitry. A supervised alarm is one that regularly checks its sensors to confirm that they're communicating properly with the control panel. With this capability, you know immediately if a sensor is broken, if the alarm's battery is low, or if a protected door or window has been left opened.

A self-contained alarm (see Figure 5–5) is a single unit that is usually shaped like a VCR; its siren, motion detector, and other sensors are

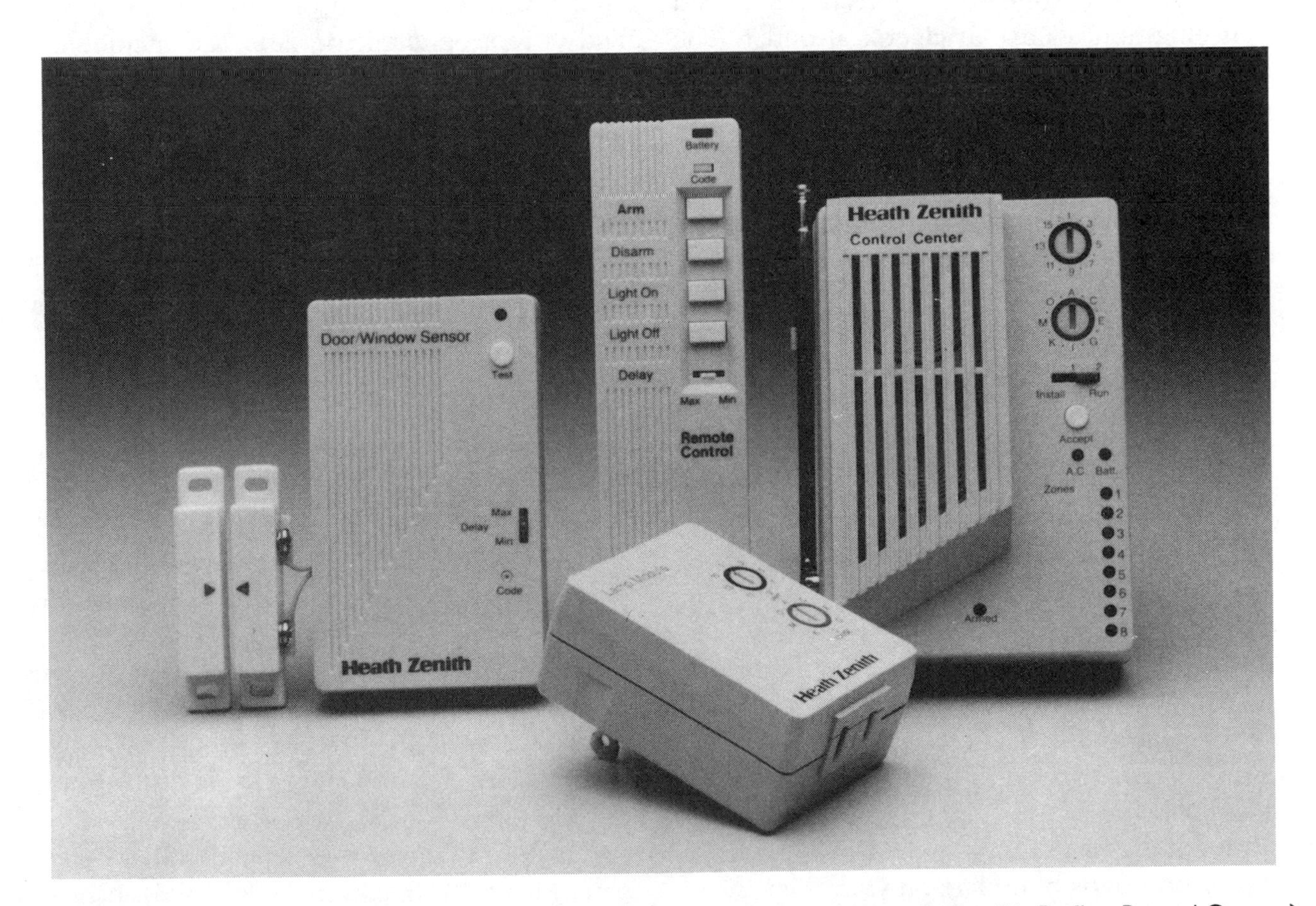

Figure 5–4. A wireless home alarm is easy to install. (Courtesy of Heath Zenith Reflex Brand Group)

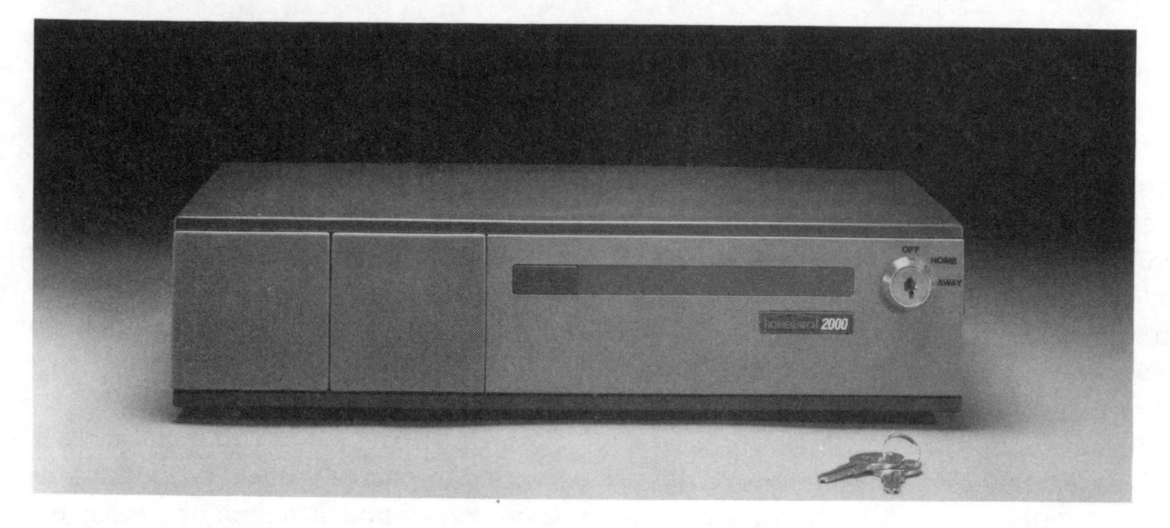

Figure 5–5. A self-contained home alarm has built-in detection devices.

all built-in. Installation is just a matter of positioning it on a sturdy table or shelf in a way that allows the unit to protect a selected area, and then plugging it into an electrical outlet. The unit sounds an alarm when someone enters the protected area.

Two advantages of a self-contained unit, compared to other types of home alarms, are ease of installation and portability. You can take the unit with you when you move or when you're traveling. Its main disadvantage is that it protects only one area at a time.

Some self-contained units can be wired to use door and window sensors, glass-break sensors, an external horn, and other accessories. Those features basically allow you to turn the self-contained unit into a hardwired system, which gives you tremendous flexibility.

What You Can Do That the Pros Won't

A seasoned professional can creatively install an alarm system so that it can thwart even the most sophisticated burglars. However, creativity doesn't come cheap. Unless you're willing to pay thousands of dollars on your home security, a professional installer isn't likely to do much more than you may be able to do for yourself. In a typical home installation, the installer will place a sensor on each entry door and a motion detector near the main entry door. (That service can cost you between $500 and $2,000!)

Because many burglars know its shortcomings, a typical professional installation may actually *increase* your risk of being burglarized. When sophisticated burglars see certain alarm system stickers on homes, they know many ways they can get in, move about inside the homes, and get out undetected. By installing your own system, you can not only save money on installation, but also make sure that all entry points are adequately protected. (Tip: Never use the window stickers that come with your alarm; they give away too much information about your system. Buy no-name window stickers from a home improvement center or hardware store.)

Most manufacturers of do-it-yourself alarms try to make it easy for you to custom-install their systems by providing heavily illustrated installation manuals, free installation videos, and technical advice by telephone. Some companies have toll-free numbers solely for providing installation and operating guidance.

How to Install a Wireless Alarm System

These instructions can be used to install many popular brands of wireless alarms, including Heath Zenith and X-10, that use a plug-in Control Center, a Remote Control, and a Lamp Module:

1. Set the Control Center's "Install/Run" switch to "Install." Place a 9-volt alkaline battery in the battery compartment. Plug the Control Center into an unswitched AC outlet. Choose a location that is central to all the door/window sensors that you plan to use. Keep in mind that you will want the Control Center where you can easily see it each day, so that you will notice whether any of the zone indicators is reporting a problem.
2. Install another 9-volt battery in the Remote Control's battery compartment. Press the code button on the Remote Control with the point of a pencil (or a straightened-out paper clip). Press the "Arm" button. The Control Center will chime once, letting you know that the Remote Control is installed.
3. Install a third 9-volt battery in the door/window sensor's battery compartment. Press the code button on the door/window sensor with the point of a pencil. Press the test button on the door/window sensor. The Control Center will chime once and assign the door/window sensor to a zone. The corresponding zone number (Zone 1, for example) will light on the Control Center.
4. Make sure that the house and unit codes on the Lamp Module are the same as the codes on the Control Center. Then plug a lamp into the Lamp Module and plug the Lamp Module into an active outlet. You should now be able to control the lamp with your Remote Control.
5. Using the supplied double-sided tape or screws, mount a door/window sensor on the wall next to the door or window you want to protect. Avoid mounting the door/window sensor on metal surfaces or walls. If you're mounting the magnets on metal doors or frames, the magnets should be no more than 3/16 inch apart. On wooden surfaces, they may be up to 3/8 inch apart. When mounting the magnets directly onto a steel frame, use a wood or plastic spacer to raise the magnets slightly off the frame.

Tricks of Hardwiring

Although hardwired systems are generally more reliable and less expensive than their wireless counterparts, few laypersons like to install hardwired alarms. Sometimes, getting a length of wire from a control panel to the sensors can be tricky. Here are some tips that might help (see Figure 5–6):

- When running wire from one floor to another, try using the openings used by plumbing or vents.
- If you have to drill a hole to get wire from one floor to another, consider drilling in a closet or another place that won't be noticeable. As a last resort, consider drilling as close to a corner as possible.
- Try running wire above drop ceilings.
- Try running wire under wall-to-wall carpet as close to the walls as possible (not in high-traffic pathways).
- If you can't hide the wire you're running, consider running it through plastic strips of conduit (see Figure 5–7). (Conduit not only makes the run look neater, but also protects the wire.)
- If you can't hide the wire and aren't using conduit, try to run the wire close to the baseboard.
- When running wire without conduit, you may need to staple the wire. Use rounded staples only; flat-back staples may cut into the wire and cause problems.

HOME AUTOMATION SYSTEMS

Although locks, light, sound, and other elements play a part in home security and safety, each of those elements must be controlled separately in most homes. By using a home automation system, however, you can make several or all of the systems and devices in your home work automatically to provide more security, safety, and convenience.

"Home automation" is a generic term that refers to any automated technology used in homes—a sprinkler system that shuts itself off during a rainfall, automatic lights that come on when someone pulls into a driveway, and so on. If the right attachments are used, all home automation systems can perform many of the same functions. However, there are important differences among the three basic types of systems:

1. X-10 compatible modules;
2. Programmable controller;
3. "Smart house" integrated system.

X-10 Compatible Modules

The simplest and least expensive home automation systems use X-10 compatible modules that need only to be plugged into any standard wall outlet (see Figure 5–8). Each module is designed to control a specific appliance. One module may allow you to turn a light on and off from a remote location; another may automatically turn on your coffee pot at a certain time.

Because the modules are sold separately, you can use them to custom-design your home automation plan (see Figure 5–9). X-10 modules are so popular that hundreds of major manufacturers of security-related products advertise their products as being "X-10 compatible." Virtually any home automation system that you can buy for under $500 uses X-10 components.

The main weakness of an X-10 system is that the components don't share memory; in effect, each component works independently of all the other components. However, some systems can be designed to coordinate the functions of several X-10 components, which gives them the

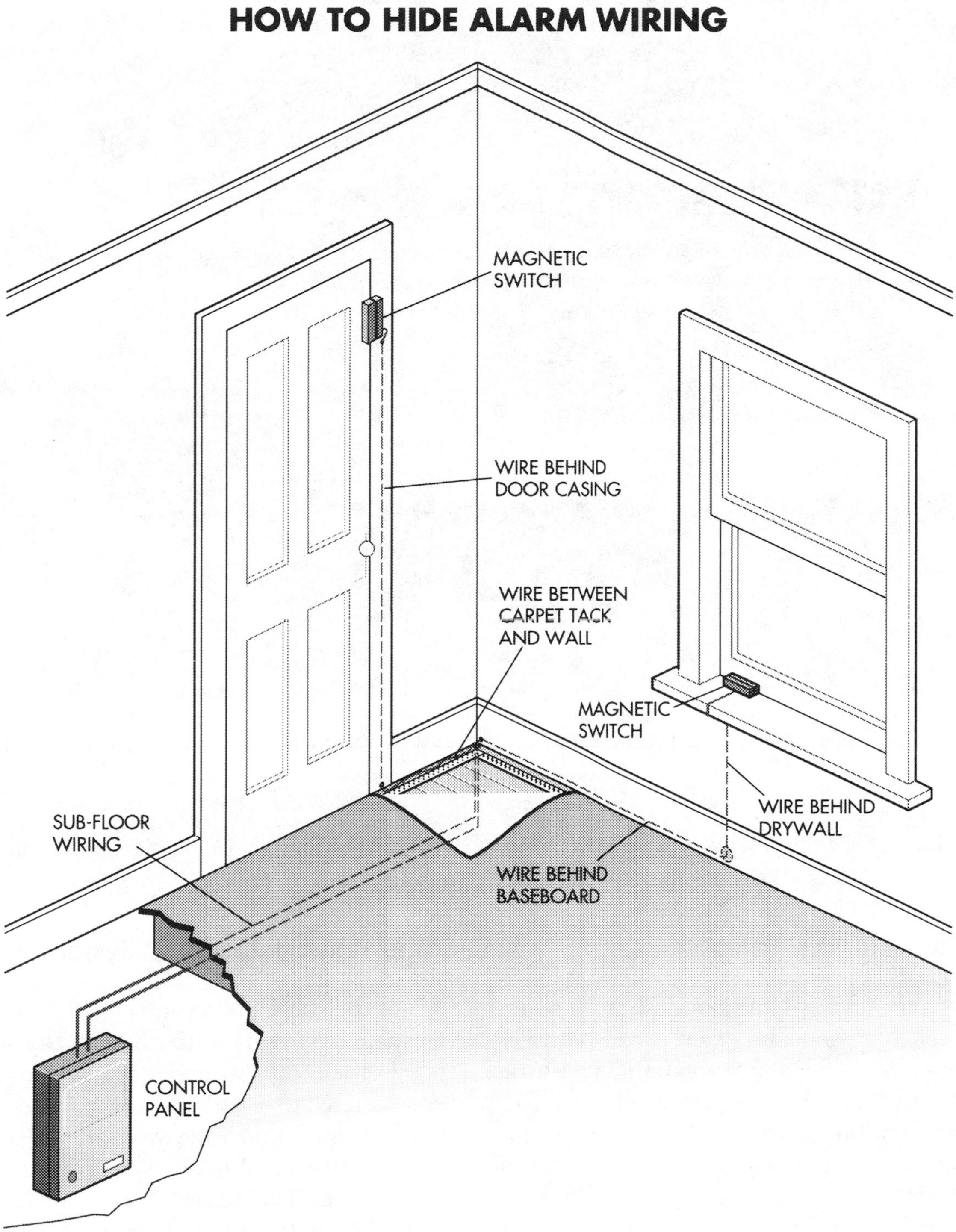

HOW TO HIDE ALARM WIRING

Figure 5–6. When hardwiring an alarm system, you need to find ways to hide as much wire as possible.

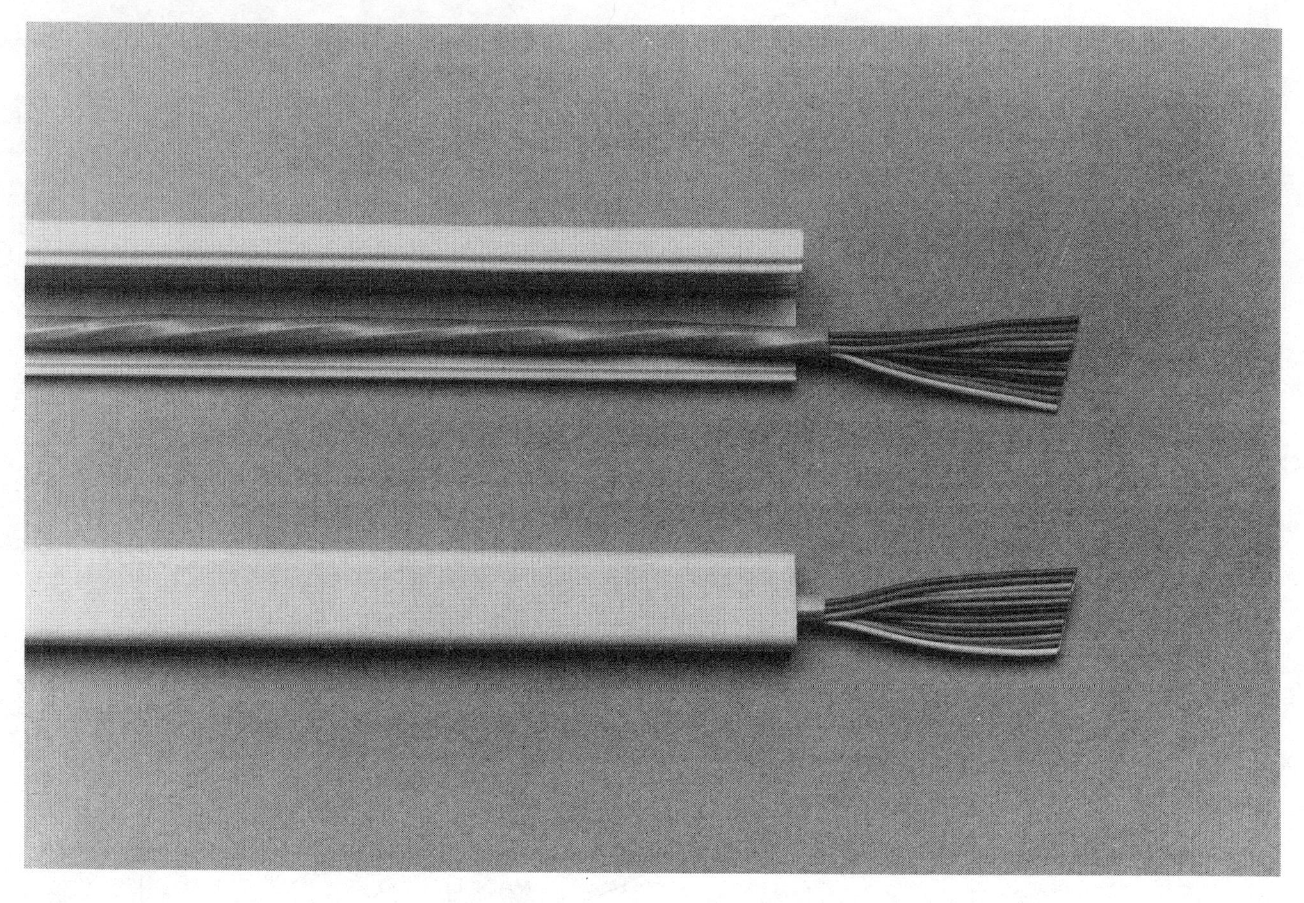

Figure 5–7. Self-adhesive conduit allows you to run wire neatly throughout a room. (Courtesy of Alarm Accessory Ltd.)

appearance of sharing memory. (To learn more about using X-10 components, see Appendix A.)

Programmable Controller

A more versatile type of home automation system is one that uses a programmable controller and is integrated into your home's electric power line. This type of system allows all of your automation devices to work together under a central control. By touching a keypad in your bedroom, for instance, you could turn down the heat in your home, arm your burglar alarm, and turn on your outdoor lights; or you could use your programmable controller to make all of those things occur automatically every night at a certain time. It can cost up to $20,000 to have a full-blown power-line system installed in a home.

Smart House Integrated System

One of the latest and most sophisticated home automation systems is the Smart House. Although the term "Smart House" is sometimes used to refer to a wide variety or a combination of home automation systems, it's actually a brand name for a unique system of automating a home. The Smart House integrates a unique wiring system and computer chip language to allow all the televisions, telephones, heating systems, security systems, and appliances in a home to communicate with each other and work together.

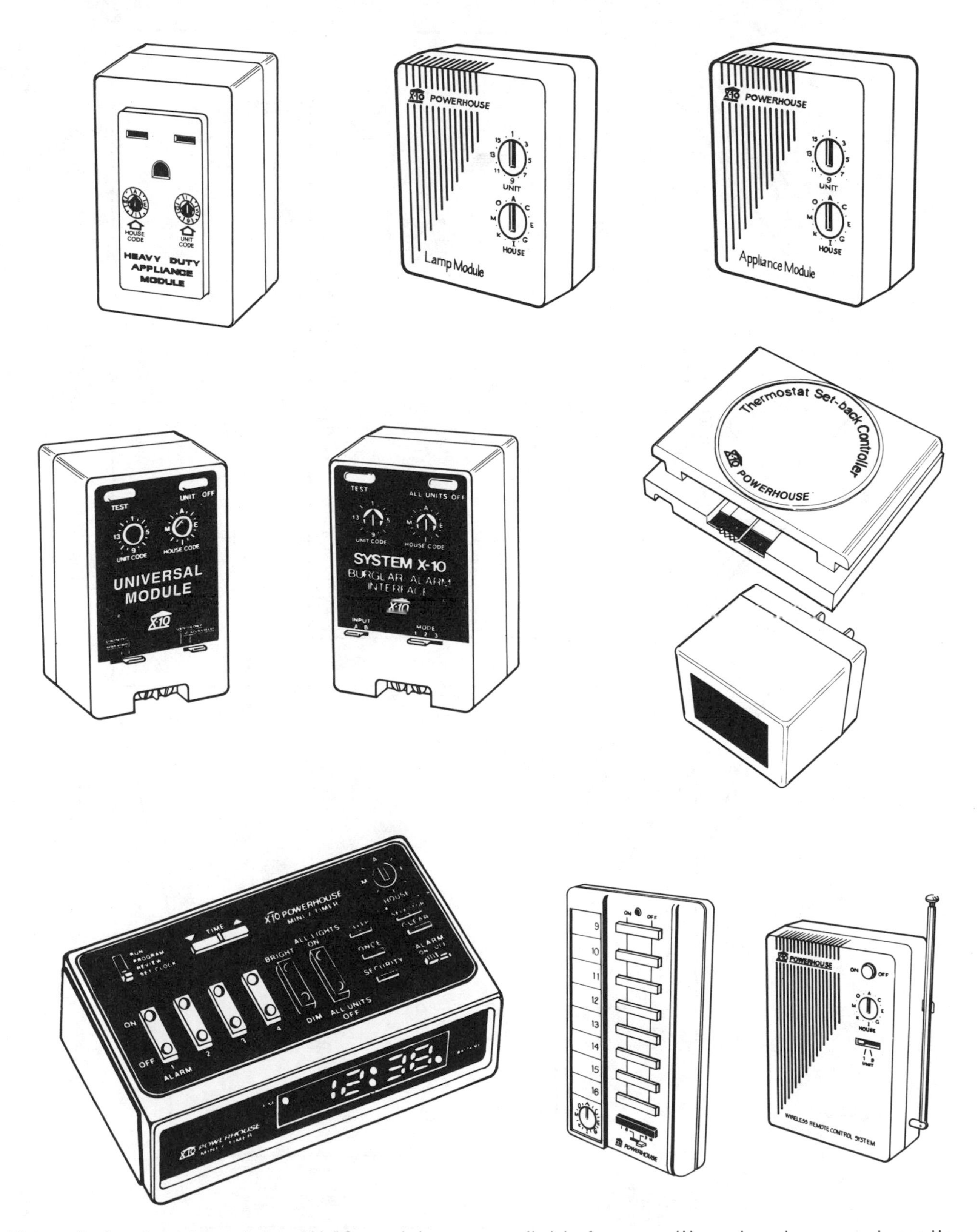

Figure 5–8. A wide variety of X-10 modules are available for use with various home automation and security systems. (Courtesy of X-10 (USA) Inc.)

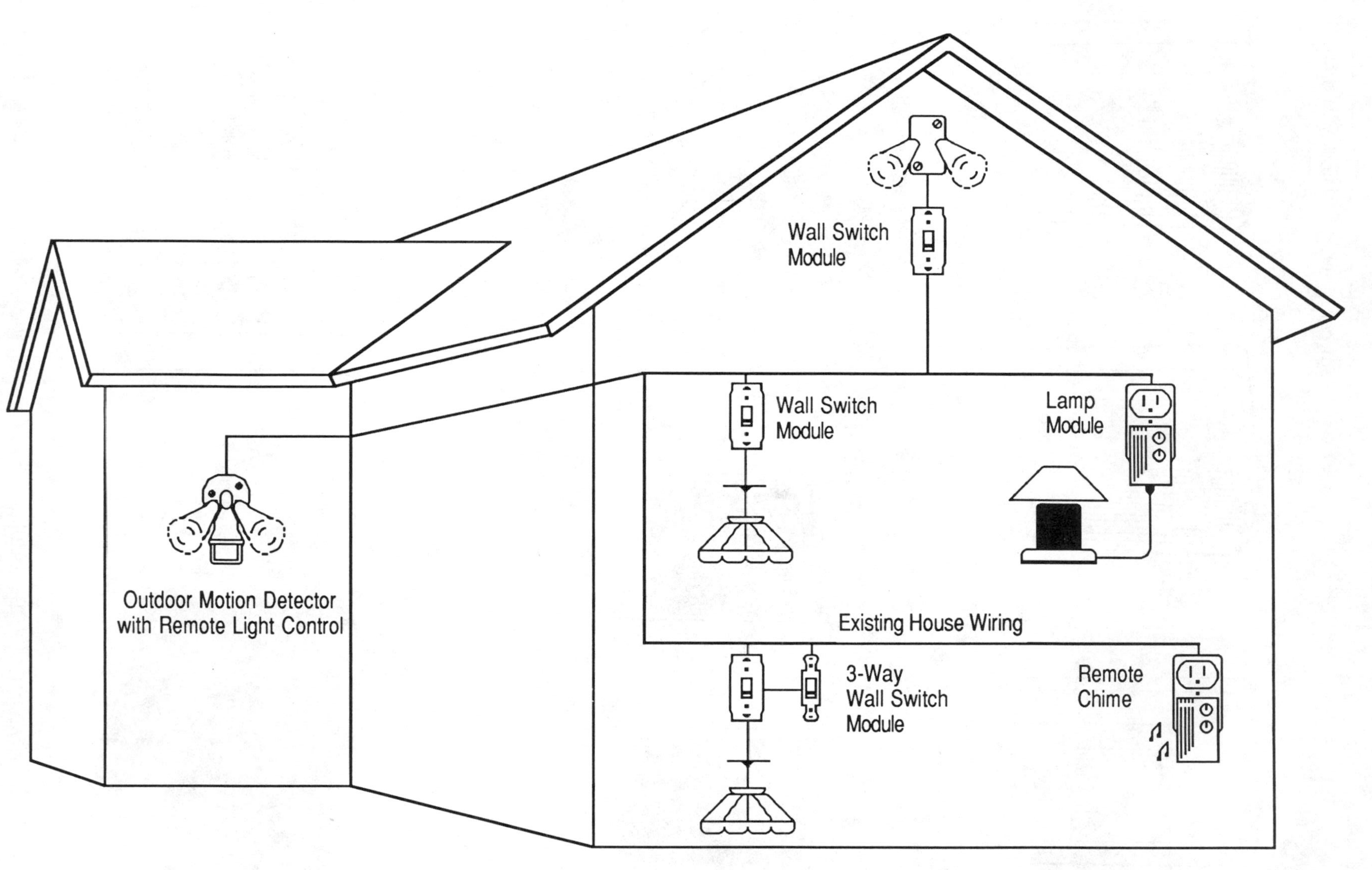

Figure 5–9. X-10 modules can be used in various ways throughout a home. (Courtesy of X-10 (USA) Inc.)

If your refrigerator has been left ajar, for instance, the Smart House could signal your television set to show a picture of a refrigerator in the corner of the screen until you close the door. A smoke detector in the Smart House could signal your heating system to shut down during a fire. Its communication ability is one of the most important differences between the Smart House and all other home automation systems. The basic installation cost of a Smart House system is about the same as that of a power-line system, but with a Smart House system you may need to purchase special appliances.

How the Smart House System Works. To understand how Smart House technology works, it's important to realize that the technology was the result of a joint effort among many appliance manufacturers, security system manufacturers, and home building and electronics trade associations. All of them agreed on standards that allow special appliances and devices to work in any Smart House.

A Smart House uses a system controller instead of a fuse panel, and "Smart blocks" instead of standard electrical outlets. Appliances that are designed to work in a Smart House are called "Smart appliances"; all of them can be plugged into any Smart block. The same Smart block into which you plug your television, for instance, can be used for your telephone or coffee pot. When a Smart appliance is plugged into a Smart block, the system controller receives a code to release power and it coordinates communication between that appliance and the other Smart appliances.

A big difference between standard electrical outlets and Smart blocks is that electricity is always present in the outlets. If you were to stick a metal pin into one of your standard outlets, you would get an electric shock. If you were to stick a pin into a Smart block, you wouldn't get shocked because no electricity would be present. Only a device that has a special computer chip code can signal the Smart House system controller to release electricity to a particular Smart block—unless you override the signal.

With a Smart House, you have the option of programming any or all of the Smart blocks to override their need for a code. That option allows you to use standard appliances in your Smart blocks in much the same way that you use your electrical outlets now. Standard appliances can't communicate with each other or with Smart appliances. You might want to override a Smart block if some of your appliances aren't Smart appliances.

Because the Smart House is a new technology, very few Smart appliances are available. As the technology becomes more widely used, the demand for Smart appliances will increase.

Home Automation Controllers

With either a Smart House or a power-line system, you need only one controller to make the system do anything you want it to do. For convenience, however, you might want controllers installed at several locations in your home. In addition to a keypad, you can use your telephone, a computer, or a touch screen for remote control of your system (see Figure 5–10).

A touch screen looks like a large television that is mounted into a wall. It displays a "menu" of your options—lighting, security, audio, video, temperature controls, and so on—and you can make your selection just by touching the screen. If you were to touch "security," for example, a blow-up of the floor plan of your home would appear on the screen and you would be able to see whether any windows or doors are open, whether your alarm system is on or off, and other conditions related to your home's security. You would also be able to secure various areas of your home just by touching the screen.

Figure 5–10. A touch screen can be used to control heating, cooling, security systems, lights, and appliances throughout a home. (Courtesy of Unity Systems)

RECOMMENDED RESOURCES

(See HM Sourcelist for more information.)

Ademco—Alarm control panels and sensors; professional alarms.

Alarm Accessory Ltd.—Installation supplies

Arrow Fastener Company, Inc.—Installation supplies

Dicon Systems Ltd.—Do-it-yourself alarms

Dimango Products Corporation—Do-it-yourself alarms

Diversified Mfg. & Mktg. Company, Inc.—Installation supplies

Eastman Wire & Cable Company—Installation supplies

Group Three Technologies—Home automation systems

Heath Zenith Reflex Brand Group—Do-it-yourself alarms

Home Automation, Inc.—Home automation systems

Honeywell, Inc.—Home automation systems

Linear—Alarm sensors

Napco Security Systems—Alarm control panels and sensors; professional alarms

Transcience—Do-it-yourself alarms

Unity Systems, Inc.—Home automation systems

Vantage Technologies, Inc.—Do-it-yourself alarms

X-10 (USA) Inc.—Alarm sensors; do-it-yourself alarms; home automation systems (See Appendix A for information on using X-10 modules, which are flexible enough to allow you to upgrade or custom-design a system.)

EXPERT ADVICE

Alexander Murray

I asked home alarm systems expert Alexander Murray to share some of his secrets for choosing and installing burglar alarms. Murray is a former electronic security instructor at the National School of Locksmithing and Alarms (New York City branch), and is a technical consultant to a major manufacturer of alarm systems.

Q. Do you think average homeowners should attempt to install their own burglar alarms, or would you recommend using a professional installer?

A. As you know, the professional has some big advantages over most homeowners—such as experience, special tools, and access to special parts and services. Those advantages can be important for commercial installations, but aren't so important to most home installations. Anyone who can use common hand and power tools should have little trouble installing an effective burglar alarm at home.

Q. Which type of alarm would you recommend to the do-it-yourselfer?

A. I love hardwired systems because I think they're the most reliable. And they're easier to troubleshoot. But there are a lot of good wireless systems available for anyone who doesn't want to run wire.

Q. Which detection devices do you prefer for homes?

A. I like foil on windows because I like the way it looks and because it acts as a deterrent. Burglars can see it from far away and know that you have an alarm system. Two other detection devices I highly recommend are magnetic switches and PIRs. I find that the PIR is the best type of motion detector for most homes. It consumes little power and offers good resistance to false alarms.

Q. I know you've installed a lot of wireless burglar alarms in homes. What procedure do you follow during a typical installation?

A. In most cases, first I choose a location for the control panel—keeping in mind that it must be protected and be able to receive radio signals. The best spot in a home is often a centrally located first-floor closet.

I make sure no large metal object—like a refrigerator—is near the control panel. Sometimes a metal object can act as an antenna and screw up the reception.

After mounting the control panel, I install the siren and keypad. I try to put the keypad as close to the control panel as possible. If I had put the control panel in a closet, for instance, I'd put the keypad right outside the closet on a wall. And whenever possible, I install the siren in the attic.

Q. In what ways does your installation procedure differ when you're installing a hardwired alarm?

A. I still first choose a location for my control panel. But with a hardwired system, I generally prefer someplace in the basement near an electrical outlet. My main concern is that the place is accessible to running wire throughout the house.

And, of course, with a hardwired system, I try to hide as much wire as possible. I like using a hooked piece of wire to snake wire between walls. But a person who hasn't ever snaked wire might have trouble doing so.

Q. Do you think homeowners need to be concerned about a burglar cutting their telephone wires to prevent their alarms from calling for help?

A. Until recently, I didn't think most homeowners needed to worry about that. But these days, more and more burglars are cutting phone lines before breaking into homes. Although there are a lot of ways to protect against that, I think the simplest thing for a homeowner to do is to install a line fault monitor in the control panel. That way, at least the siren will sound as soon as the phone line is cut.

Q. In what situations do you think it's better for a homeowner to use a professional alarm installer? And how would you suggest finding a reputable installer?

A. The more doors, windows, skylights, and other points of entry a home has, the more likely a professional installer can save the homeowner time and money. The best way to find a good alarm installer is to talk with friends and neighbors who have used one. Personal recommendations mean a lot more than telephone book advertisements. And, a homeowner would be wise not to ignore small companies; they often offer all the services a homeowner needs at lower prices than big companies.

Tips for Working with Electricity Safely

Although working with electricity can be very dangerous, it doesn't have to be, if you have respect for its power and always take proper security precautions. Here are ten important rules to follow.

1. Before working with an electrical circuit, always shut off the power at the fuse box or circuit breaker.
2. Post a sign on the fuse box or circuit breaker to let people know that you're working on one of the circuits and that they shouldn't turn the power back on.
3. Use a voltage tester to make sure the power is off before you touch any wires.
4. Always unplug a device before working on it.
5. Make sure your hands are dry before installing or servicing electrical equipment.
6. Never stand on a wet floor while working with electricity.
7. Don't touch plumbing or gas pipes while working with electricity. If you touch such a pipe and a hot wire at the same time, you could be badly shocked.
8. Don't run an extension cord across a doorway or other place where someone might trip over it.
9. Never plug in equipment that has a frayed or damaged cord. Repair or replace the cord.
10. When removing a cord from an outlet, pull on the plug, not on the cord.

6

Lighting for Security, Safety, and Beauty

Lighting is not only a low-cost form of security; it can also help to prevent accidents, create moods, and enhance the beauty of any home. This chapter shows how you can make the best use of lighting inside and outside your home. A dark house is an invitation to crime and creates a high risk for accidents. When you approach your home late at night, you need to be able to walk to your entrance without tripping over something—or *someone*—in your path. When you're inside, you need to be able to move from room to room safely. Your home should be well-lighted on the inside, in the areas directly outside the doors, and throughout the yard (see Figure 6–1). "Well-lighted" doesn't necessarily mean a lot of light; it means having the light sources and controls in the right places.

Figure 6–1. Landscape lighting can help prevent accidents, deter burglars, and enhance the beauty of a yard. (Courtesy of Progress Lighting)

LIGHT SOURCES

Our most common light source is the sun, which we cannot control. We have artificial light sources available for use at night and in some indoor locations during the day. Important differences among artificial sources include color, softness, brightness, energy efficiency, and initial cost.

Light sources you might consider for home use include standard incandescent, halogen, fluorescent, and high-intensity discharge (HID) lighting. The HID family of lighting includes low-pressure sodium, high-pressure sodium, mercury vapor, and metal halide.

An "incandescent" light source relies on heat to produce light. The standard bulbs used in most homes are incandescent (lighting designers call them "A-lamps"). They have a metal filament that is heated by electricity. Standard incandescent bulbs are inexpensive, readily available, and suitable for most home fixtures. They light up almost immediately at the flip of a light switch. However, using heat to produce light isn't energy-efficient; in the long run, incandescent lighting can be more costly than other sources that require special fixtures.

Halogen, a special type of incandescent source, is slightly more energy-efficient than standard incandescent lighting. A halogen bulb uses a tungsten filament and is filled with a halogen gas.

Fluorescent lighting uses electrical current to make a specially shaped (usually tubular) bulb glow. The bulbs come in various lengths, from 5 inches to about 96 inches, and they require special fixtures. You might not want to use fluorescent lighting with certain types of electronic security systems, because it can interfere with radio reception. Nor would you ordinarily use fluorescent lighting outdoors in cold climates: it's very temperature-sensitive.

For outdoor lighting, you might use high-intensity discharge (HID) sources, which are energy-efficient and cost little to run for long periods of time. Like fluorescent lighting, HID sources require special fixtures and can be expensive initially. Another potential problem with HID sources is that they can take a long time to produce light after you've turned them on. Start-up time can be unimportant if you use a light controller to automatically activate the lights when necessary.

LIGHT CONTROLLERS

Timers are among the most popular types of controllers for indoor and outdoor lighting. The newest timers can do much more than just turn lights on and off at preset times (see Figure 6–2).

Programmable 24-hour wall switch timers, for example, will randomly turn your lights off and on throughout the night and early morning. That feature is useful because many burglars will watch a home to see whether the lights come on at the exact same time each night—an indication that the home is empty and a timer has been preset. Another feature of some new timers is built-in protection against memory loss. After a power failure, they "remember" how you had programmed them. Some models adjust themselves automatically to take into account Daylight Savings Time changes. Most timers used by homeowners cost between $10 and $40.

Another low-cost way to control lights is with sound or motion sensors. You can buy one of these sensors and connect it to, say, a table lamp in your living room (see Figure 6–3). When you (or someone else) walk into your living room at night, the light will come on automatically.

Some floodlights come with a built-in motion sensor (see Figure 6–4). If you install them outside at strategic places, they will warn you of nighttime visitors. You might install one

Figure 6–2. Some modern timers combine sophisticated programming, digital read-out, and plug-in convenience. (Courtesy of Intermatic, Inc.)

facing toward your driveway, for instance, so it will light up the area when a car pulls in. Floodlights generally sell for less than $50.

PREVENTING ACCIDENTS

To prevent accidents at nighttime, you need to be able to see potential hazards. When walking down a flight of stairs, for instance, it's important to be able to see whether any objects are in your way. In too many homes, there is a need to stumble through dark areas or to grope for a light switch or a series of light switches, before reaching the bathroom or kitchen.

Can you use motion-activated sensors to avoid that problem? You'd probably need a lot of them to cover every path you might take at night. Simpler and more convenient options are available.

One useful practice is to install night lights near your light switches, so you can more easily reach them. Night lights cost only a few dollars each, and they consume very little power.

Another option is to use three-way switches. They allow you to turn a light on and off at more than one location, such as at the top and bottom of a flight of stairs.

The most convenient way to use lighting indoors is to tie all (or most) of the switches into an easily accessible master control panel. You could then turn on a specific group of lights by just pushing a button. One button could turn on a pathway of lights from, say, your bedroom to the bathroom. By pushing another button, you could activate a pathway of lights from your bedroom to the kitchen. Some or all of your outdoor lights could also be tied into your master control system.

Installing Motion-Activated Outdoor Lighting

You can install a motion-activated light virtually anywhere indoors or outdoors—on the side of your home, on your porch, in your garage, or wherever it's needed. Many of these lights are simple two-wire installations in which the hardest part of the process is the proper positioning of the lighting unit. Models are available in various colors and styles to match your decor (see Figures 6–5, 6–6, and 6–7).

Figure 6–3. When connected to a lamp, a motion sensor will turn the light on automatically whenever someone walks near it. (Courtesy of Falcon Eye, Inc.)

120-Volt Lighting

A 120-volt lighting system will provide brighter light than a low-voltage system. The brighter light may be especially useful outside, if you need to illuminate a large area. Installing a 120-volt system is more involved and the materials are more expensive than those used in a low-voltage system.

Before beginning the installation, familiarize yourself with your local electrical code and obtain any required permits. You may have to draw up a plan and have it reviewed by your local building inspector before you're allowed to install 120-volt lighting.

You'll need to decide what materials to use—receptacles, cables, switches, boxes, conduit, conduit fittings, wire connectors, and so on. Your local code may have already made some of those decisions for you. You may be required to use rigid metal conduit rather than PVC conduit, for instance, or you may be restricted to using only certain types of wire.

Figure 6–4. Some floodlights come pre-assembled with a motion detector. (Courtesy of Heath Zenith Reflex Brand Group)

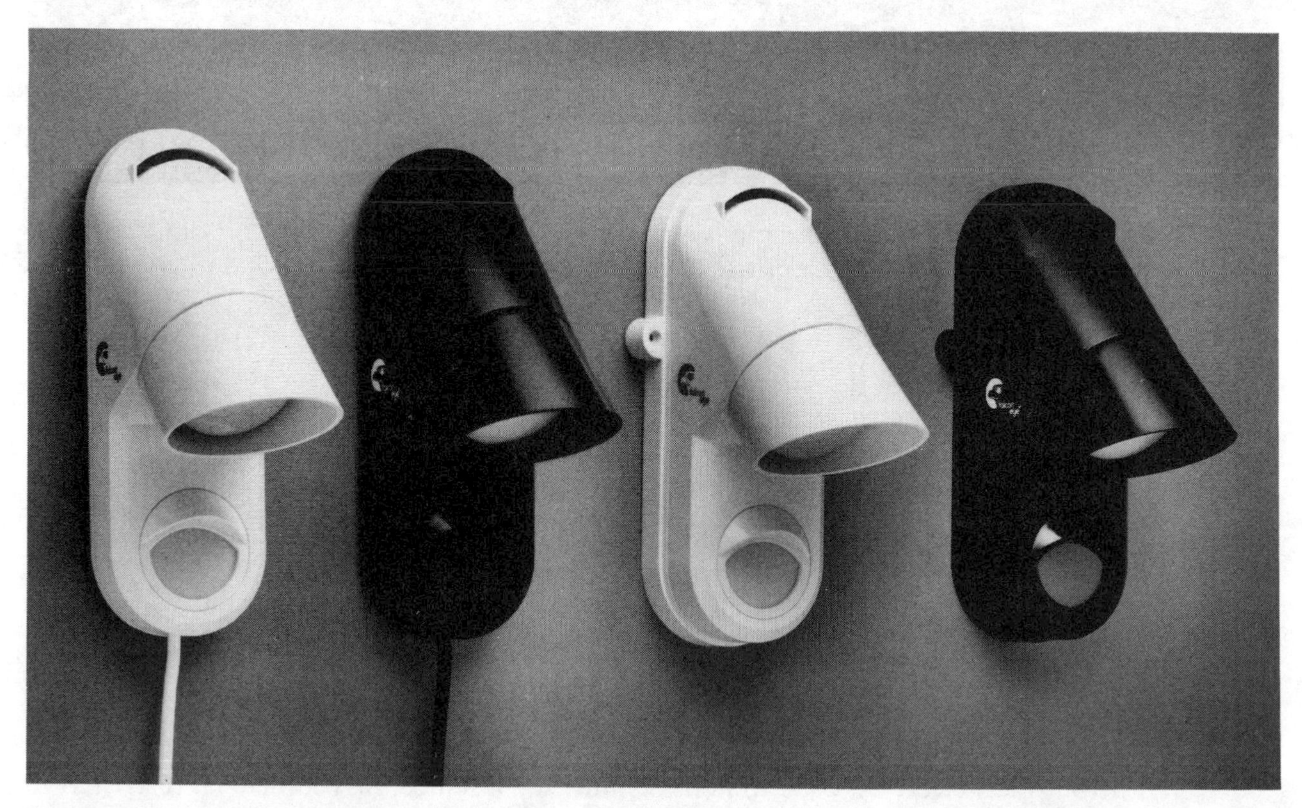

Figure 6–5. Outdoor lights with built-in motion sensors come in a variety of styles and colors. (Courtesy of Falcon Eye, Inc.)

THINGS YOU SHOULD KNOW

The unit should be mounted 6 feet to 9 feet high – the higher the unit is mounted, the larger the detection zone and lighted area.

A built-in photo cell prevents the unit from operating during normal daylight hours. This also means that bright lights or reflections from light colored surfaces may keep the unit from turning on.

By detecting differences in heat along with movement, your Falcon Eye automatically turns itself on the instant anyone enters the detection zone. Because differences in heat is one of the triggering factors, the unit may be less sensitive and react more slowly on unusually warm nights – when air temperature and body temperature are relatively close.

The light will automatically turn itself off approximately 5 minutes after detection zone is vacated.

Installation Instructions for Plug-in Models

1. Use template printed at left to position and install the 2 mounting screws enclosed. (Also enclosed are 2 plastic inserts for mounting on masonry, plaster or sheetrock walls.)

2. Screw bulb into fixture.

3. Hang unit on mounting screws.

4. Plug into electrical outlet. If controlled by a light switch, make sure the switch is in the ON position.

Your light will usually come on and then cycle off (daylight or dark) when it first receives power.

For added security, you may want to place a screw or nail at the top of the installed unit to make removal more difficult.

Figure 6–6. Tips for mounting a plug-in model Falcon Eye motion-sensing light. (Courtesy of Falcon Eye, Inc.)

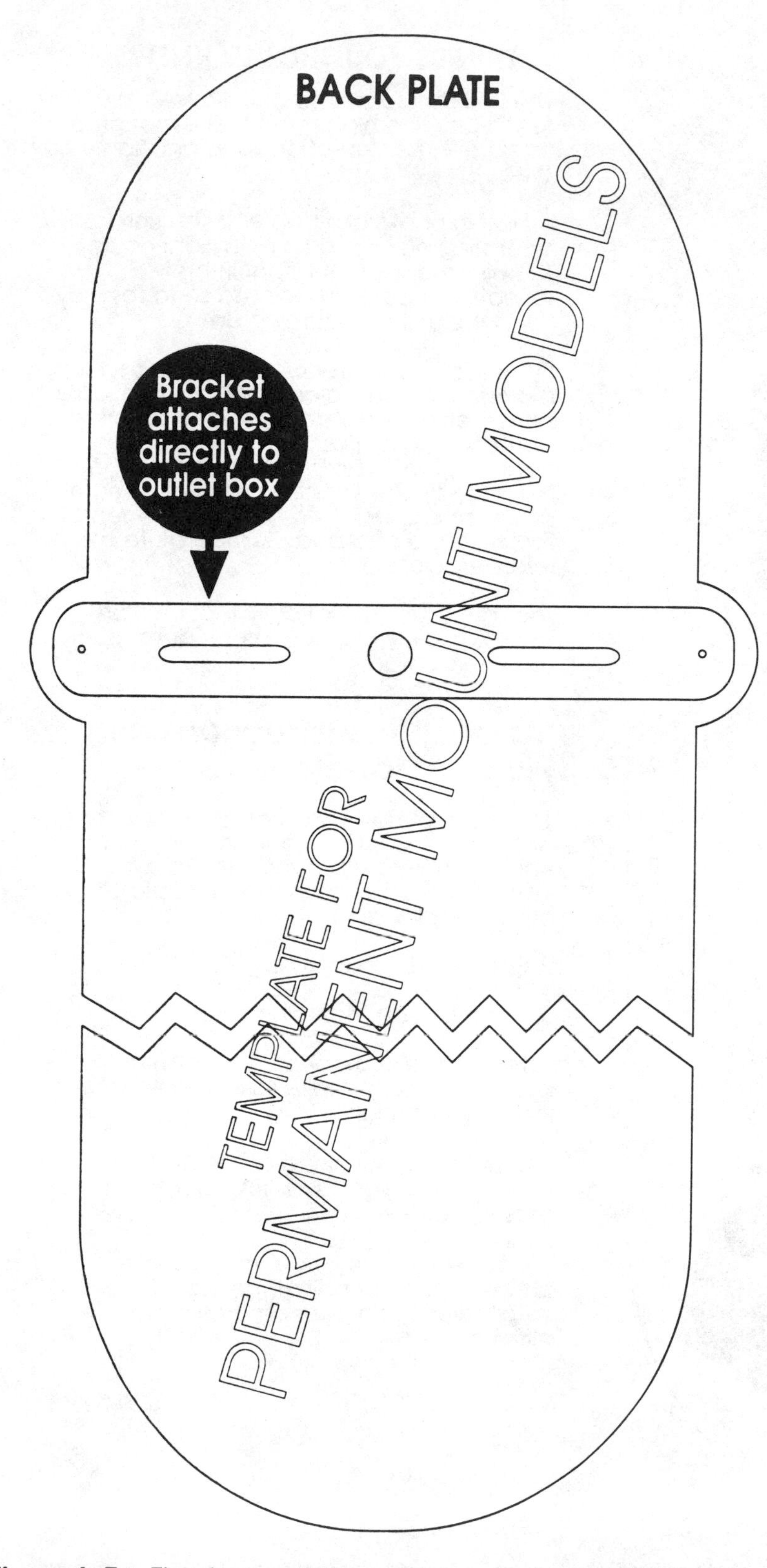

Installation Instructions for Permanent Mount Models

1. Turn off electrical power before starting installation.

2. Using the 2 screws enclosed, attach bracket to electrical outlet box with flanges facing out – away from wall.

3. Using the wire nuts provided attach the black wire coming out of the back of the Falcon Eye to the black (positive) wire from the outlet box and the white wire to the white (negative) wire.

4. Turn power back on. Your light will usually come on and then cycle off (daylight or dark) when it first receives power.

Figure 6–7. Tips for mounting a permanent mount Falcon Eye motion-sensing light. (Courtesy of Falcon Eye, Inc.)

EXPERT ADVICE

Julia Rezek

I asked lighting designer Julia Rezek for her secrets to choosing and installing residential lighting. She began working as a lighting designer in 1983, after graduating from UCLA. Ms. Rezek has taught lighting classes at UCLA's University Extension Program and at the Otis Parson Institute of Design in Los Angeles. She is the Director of Residential Lighting for Grenald Associates in Culver City, California, and is a member of the International Association of Lighting Designers.

Q. What suggestions would you give a homeowner who wants to use lighting for security but doesn't want the lighting to be harsh and offensive?

A. I recommend installing two complementary types of lighting systems. One is for panic situations—such as when you're awakened by a noise at night—to turn on very bright and offensive floodlights in your yard or driveway. The other lighting system should be much softer and be connected to a time clock so it can come on automatically every night.

Q. Which lighting sources and fixtures do you recommend for indoor lighting?

A. When planning the lighting for an indoor space, I consider what activity the space will be used for, what effect is desired from the light source, and how the lighting should be controlled. A primary concern for indoor lighting is to avoid glare and create an environment that is soft, comfortable, and easy to control.

Most homeowners tend to use only standard incandescent light bulbs, but such light is yellow and deficient in the blue-green end of the color spectrum. That's why when you're trying to select your socks in the morning you can't tell the difference between the blacks and browns and blues.

I like low-voltage halogen for displaying artwork because it's much whiter than regular incandescent and renders colors more accurately. Fluorescent lighting is more energy-efficient, but the type found in many homes produces a cool white light and does a poor job of rendering color—which is why many homeowners dislike fluorescent lighting.

But, for a few extra dollars, you can buy high-color-rendering fluorescent lighting. That type is often used in expensive health clubs because it makes a person's skin tone look great—like the person just walked out of a tanning salon.

I also like using recessed lighting for wall washing effects and accenting, and track lighting where it can be concealed. And I like some of the more contemporary sources that are integrated into the architecture—such as cold cathode or neon or linear incandescent strip lighting.

Q. Which lighting sources do you recommend that homeowners use outdoors?

A. I find that metal halide is one of the best sources for lighting landscapes at night. It renders greens, blues, reds, and yellows very well. Mercury vapor can also be a good source for landscape lighting, but it can be a little tricky to work with. The only way to be sure of which light source is best for a particular piece of property is to actually try different sources. That's what we lighting designers do.

By installing various types of HID sources in large trees, you can light up acres of land and create a wonderful sense of security. The lights will be very powerful, but will make the property look beautiful. That kind of lighting brings the nighttime environment alive.

Q. Are there outdoor lighting alternatives that you would suggest to someone who doesn't have large trees or who doesn't want to install lights in trees?

A. A person can use low-voltage lighting—like the MR16 halogen lamp and some of the smaller incandescent sources. They don't light up a lot of area, but are easy to install. They use small fixtures and don't have to be run in rigid conduit.

Q. What tips can you give on installing low-voltage lighting outdoors?

A. First, anyone installing outdoor lighting should know how to comply with all electrical codes, which vary from one city to another. A typical installation would involve installing transformer boxes somewhere in the back of the house, around the corner, or somewhere that they can't be seen. Then the homeowner would need to dig a trench about 6 inches deep to run flexible wire to the different light sources. Although it isn't usually necessary to run the wire through rigid conduit, it's a good idea to run it through PVC piping to prevent a lawn mower from damaging the wire.

Q. What lighting sources do you recommend for porches and stoops?

A. Those are transition areas where people are coming from the outside to the inside. You want those areas to have a warm and yellow quality, so your guests will feel drawn toward the warmth of light and will know that inside your home is also warm and inviting.

RECOMMENDED RESOURCES

(See HM Sourcelist for more information.)

AAMSCO Mfg., Inc.—Lighting products

American Home Lighting Institute—Trade association

GE Lighting—Lighting products

George Kovacs Lighting Inc.—Lighting products

GTE Electrical Products—Lighting products

Heath Zenith Reflex Brand Group—Timers

Idaho Wood—Outdoor lighting

Intermatic—Timers

Lightoiler, Inc.—Indoor and outdoor lighting

Niland Company—Lighting products

Philips Lighting Company—Lighting products

Progress Lighting—Indoor and outdoor lighting

Rejuvenation Lamp & Fixture Company—Indoor and outdoor lighting

Thomas Industries, Inc.—Indoor and outdoor lighting

X-10 (USA) Inc.—Timers

7

Closed Circuit Television Systems

Imagine having extra pairs of eyes that can be several places at once. Without getting out of bed, you would be able to see who's at your front door, who's walking around your yard, or who's driving onto your property. You could also keep an eye on any room of your home. With a closed circuit television (CCTV) system, you can do all those things and more.

This chapter discusses how you can use CCTV and related equipment, how to get good buys on the equipment you need, and how to install a new CCTV system.

HOW CCTV WORKS

A CCTV system simply transmits images to monitors that are connected to the system's camera. The system's basic components are a video camera and monitors connected to it by a coaxial cable. That type of installation wouldn't be very useful for security purposes: you would have to prop the camera up in a room, point it to a fixed location that you wanted to protect, and then go and stare at the monitor.

For security, you need a camera that works while you're not around, can be controlled from a remote location, and can be connected to a burglar alarm system. If you need to monitor more than one location—your front door *and* your back door—you may want to be able to use one monitor that is receiving images from both cameras. You may also want the option of monitoring both cameras at once. All of those and many other features are possible with CCTV systems that are currently selling for less than $400 (not including installation).

When you know what's available, you can choose the cameras, monitors, and optional components that will create a custom CCTV system within your budgeted amount. If the system isn't too complex, you'll probably be able to install it yourself. Most of the CCTV systems used by homeowners are easier to install than a hardwired alarm system.

Cameras

Two types of cameras are commonly used in CCTV systems: the tube camera, the older type, which is fast becoming obsolete, and the closed coupler device (CCD) camera, which lasts longer and works better. CCDs cost a little more than tube cameras, but they have been steadily coming down in price while the price of tube cameras has remained the same. With the demand for CCDs continuing to outpace the demand for tube cameras, many camera makers are discontinuing their line of tube cameras.

Cameras come in color and black-and-white transmission models. A color camera requires maximum and constant light to be able to view a scene properly, and shouldn't be used outdoors or in any area that sometimes gets dark. A black-and-white camera is more tolerant of all types of lighting conditions and is less expensive.

The choice between color and black-and-white transmission is usually simple. In virtually every residential situation, black-and-white is more cost-effective and much less troublesome. Color cameras are needed only in banks and at other surveillance sites where the cameras' videotapes may become evidence in a court case.

Cameras come in many sizes, described by their lens diameter. The three most common sizes are 1/3 inch, 1/2 inch, and 2/3 inch. The 2/3-inch camera covers more area and gives better resolution than the 1/3-inch camera. Generally, the larger the camera, the better the picture.

Monitors

In many ways, your choice of a monitor is as important as your choice of a camera. The quality of the picture you receive on your monitor depends on both. The camera and monitor work together much like speakers and an amplifier in an audio system. If you have a great amplifier with poor speakers, or great speakers with a poor amplifier, you'll get poor sound because the sound is filtered through both devices before you hear it. In a CCTV system, the picture is filtered through both the camera and the monitor before you see it.

For transmission of a color image, your monitor and your camera must be color equipment. If either is black-and-white, you'll receive a black-and-white picture. Monitors, described based on their screen diameter, range in size from about 4 inches to over 21 inches. The most common monitors for home use are the 9-inch and 12-inch sizes.

To save money, you can buy a radio frequency (RF) modulator for your television and convert it to a monitor. Then you'll be able to view the camera's visual field just by turning your television to a particular channel (usually, either channel 3 or channel 4). You can buy an RF modulator for under $50.

Peripheral Devices

One of the most popular devices for CCTV systems is a pan-and-tilt unit. It gives a camera the ability to tilt up and down and to rotate up to 360 degrees left-to-right or right-to-left. With a pan-and-tilt unit, you will be able to zero in on items (and people) within a wider camera range. By using a pan-and-tilt unit in a large installation, you'll need only one camera, not several.

Pan-and-tilt units have long been used in airports, banks, and other large commercial installations. Because the units often cost over $1,000, they're rarely included in home CCTV systems.

Another complementary unit you can use with your CCTV is a sequential switcher. With one monitor, the switcher will allow you to receive pictures from several cameras. You can watch one camera field for a while and then switch over to another.

If you want a continuous record of what the camera sees, you can install a video lapse recorder, which will span up to 999 hours with individual photo frames, on one standard VHS 120 videotape. If you wish, the current time and date can be automatically recorded on each frame.

If you want to record only unwanted persons who enter a particular area or room, you can use a camera that has a built-in motion-detecting capability and is connected to an alarm system. The alarm will be triggered when the camera begins taping.

Another option available with today's CCTVs is the dual quad unit, which gives a standard monitor the capability of showing as many as 16 pictures at one time, from 16 separate cameras. Dual quad units can cost anywhere from $1,000 to $15,000.

Installing a CCTV System

The specific installation methods that are best for you will depend on the components you've chosen and how you want to use them. Many of the hardwiring methods for installing burglar alarms, detailed in Chapter Four, are useful for installing CCTV systems.

Most CCTV systems can be either installed independently or incorporated into a burglar alarm system. If you tie the CCTV into a burglar alarm system and use a videotape recorder, you can set the CCTV to begin recording automatically whenever the alarm is triggered. You can also have your system record sounds.

Your installation can be either overt or covert. Most homeowners use an overt system because they want would-be burglars to know they are being watched. A camera that's prominently connected to the side of your home would certainly act as a deterrent.

Some people consider cameras inside a home unattractive and threatening, but there are some advantages to keeping the cameras out of sight. Hidden cameras allow you to make a secret videotape of a burglar.

To install a covert system, you'll need to buy small cameras specially designed to be installed in the corner of a wall or in a wall cutout. They sell for between $100 and $200. Some covert cameras are disguised to look like clocks and other common objects. Their prices start at about $1,000.

You should seek legal advice before installing a covert system; in some jurisdictions, such systems are illegal. Some states consider secret audiotaping to be in violation of wiretap laws.

VIDEO INTERCOMS

A video intercom system is a CCTV that lets you talk to the people you're seeing through its camera. With many models, you can choose to see and hear a person without the person knowing you're home. Like burglar alarms, video intercoms can give your home a high level of security—and they're usually easier to install than burglar alarms.

In some video intercom systems, the cameras, monitors, intercoms, and peripherals are all separate components. Other systems have integrated components, such as a monitor with a built-in intercom or a camera with built-in peripherals. (See Figure 7–1.) Because they have fewer components, integrated units are usually simpler to install. They also tend to take up less space and look nicer. The main problem with most integrated systems is that they can't be expanded to add on sophisticated peripherals.

Some models are designed to incorporate a variety of peripherals. Aiphone's Video Sentry Pan Tilt, for example, includes an integrated camera, a monitor, an intercom, and a motorized pan-and-tilt unit. Its camera can scan 122 degrees horizontally and 76 degrees vertically—up to four times the area visible with a fixed camera. A button on the monitor unit allows you to control the panning and tilting actions of

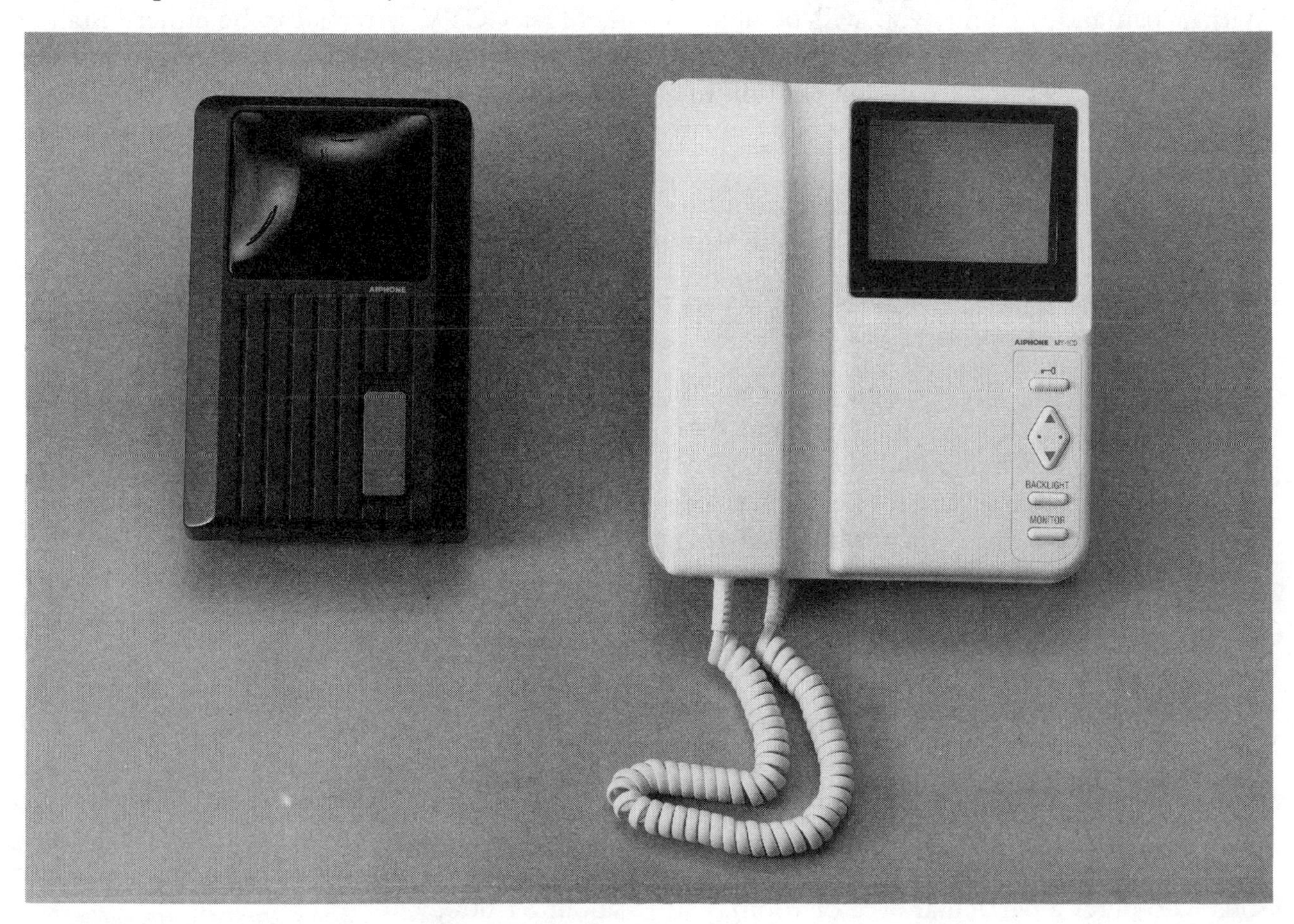

Figure 7–1. A video intercom allows you to see and talk to someone who's at your door. (Courtesy of Aiphone Corporation)

83497500 0891 Ⓓ 

VIDEO SENTRY PAN TILT

Model; MY-1CD (Room station with video monitor)

INSTRUCTIONS

NAMES AND FUNCTIONS

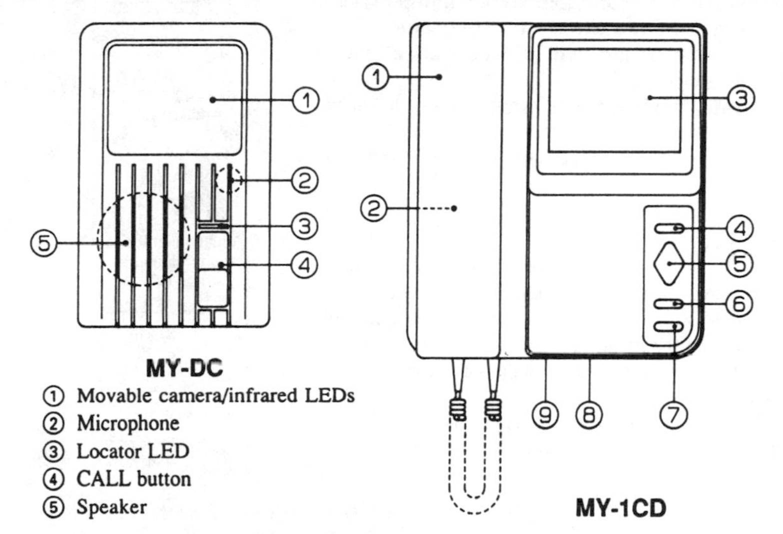

MY-DC

① Movable camera/infrared LEDs
② Microphone
③ Locator LED
④ CALL button
⑤ Speaker

MY-1CD

① Handset
② Chime volume control
③ Video monitor (4" CRT)
④ DOOR RELEASE button
⑤ PAN-TILT button
⑥ Backlight button
⑦ Video monitor button
⑧ Brightness control
⑨ Contrast switch

MOUNTING ACCESSORIES;

MY-1CD

- Stand & screw (2) (desk use)
- Wall-mounting bracket
- Screw (4) (for mounting to gang box)
- Wood screw (4) (for mounting to wall surface)

PS-18YC

- Mounting bracket (2)
- Screw (2) (for attaching bracket to unit)
- Wood screw (2) (for wall mounting)

MY-DC

- Screw (4) (for mounting bracket)
- Panel fixer
- Remover (used only for repair)

PS-18YD

- Mounting bracket
- Screw (2) (for attaching bracket to unit)
- Wood screw (2) (for wall mounting)
- Wood screw (ea. 1) (for mounting to gang box)

FEATURES

* Room station PAN-TILT button allows remote control for video door station camera vertically and horizontally, expanding the image-capturing area 4 times wider.
* BACK LIGHT button to make caller's face brighter.
* Non-polarized wiring by 2 straight wires.
* Infrared LEDs-mounted CCD camera captures image clearly even in the dark.
* Door release button.

Figure 7–2. Installation instructions for the Video Sentry Pan Tilt. (Courtesy of Aiphone Corporation)

BEFORE YOU INSTALL AND OPERATE THE EQUIPMENT - Prohibitions and precautions-

Installation;

① Do not connect any terminal on any unit to AC power lines. When you mount MY-DC/MY-1CD in place of existing bell or chime, be sure to disconect wires from the present transformer.

② Do not open the front cover of MY-1CD unit, without first removing plug of power supply from AC outlet. The high voltage is loaded on the monitor unit inside.

③ Do not drop or hit the unit. This may cause damage to the picture tube.

④ Avoid running the connecting wires through doors, windows or between furniture, which may pinch and disconnect the wires.

⑤ **IMPORTANT- Use a parallel 2-conductor cable. When using existing two wires, strictly observe that the two wires are of the same cable. Do not use any of the following wire, as they will seriously affect the quality of picture and communication.**
*** coaxial cable, * two separate and individual wires, * a pair in multi-conductor cable.**

⑥ In case noise from power supply line is induced, take terminal on MY-1CD unit to earth ground.

Installation location;

⑦ Do not install MY-DC/MY-1CD in a place where it is exposed to; *direct sunlight, *air conditioner heat, *frost, *vapor, *water, *chemicals, *iron dust, etc. OR for MY-DC, *where the strong light source is behind the person standing at the door station. The face will appear dark on the room station monitor. This situation can occur when buildings create shadows on entry areas. In this case, use BACKLIGHT button to make the face brighter.

* As shown, it is recommended that the MY-DC be installed at an entrance not exposed to direct sunlight and with 200 to 300 lux secured on the object.

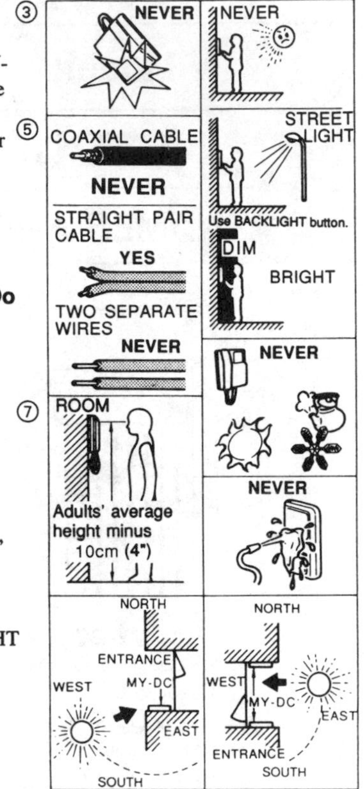

Video door station installation height & image-capturing range;

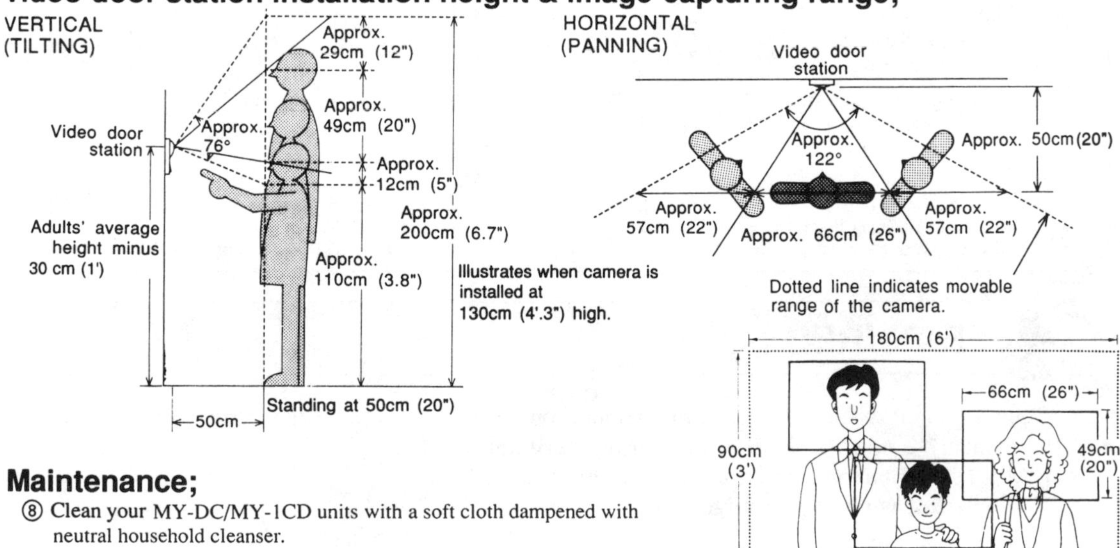

Maintenance;

⑧ Clean your MY-DC/MY-1CD units with a soft cloth dampened with neutral household cleanser.
Never use lacquer thinner or benzine, etc.

⑨ Do not splash water directly on MY-DC unit.

Figure 7–2. *Continued.*

4 INSTALLATION

(1) MY-1CD wall-mounting;

Attach the mounting bracket to SINGLE-GANG BOX (or to wall) with the supplied two screws (or wood screws). Mount the MY-1CD unit onto the bracket.

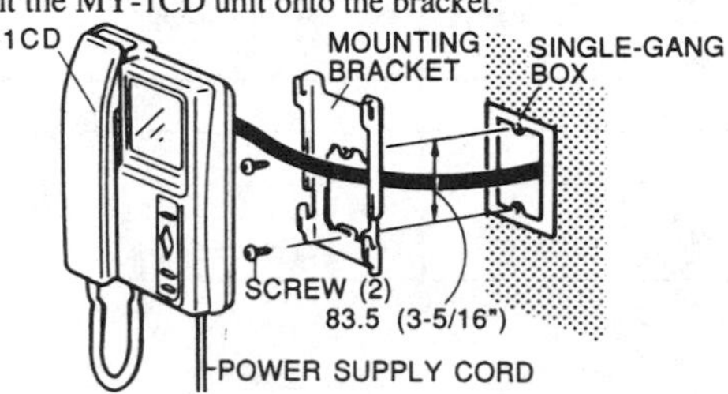

(2) Desk-top installation;

Select a location where MY-1CD unit will not fall or where the cable may cause tripping.

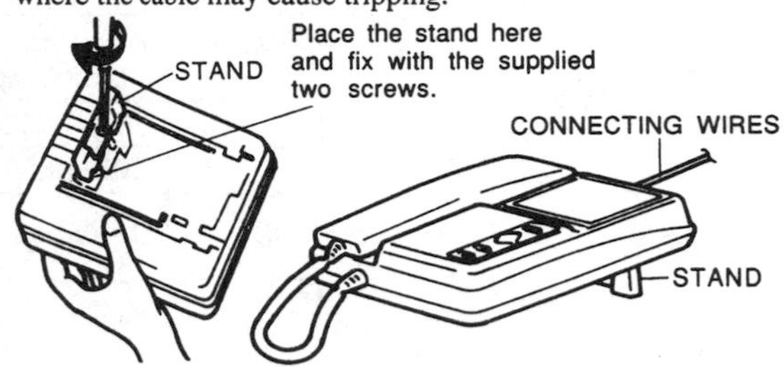

(3) MY-DC wall-mounting;

Semi-flush mounting;

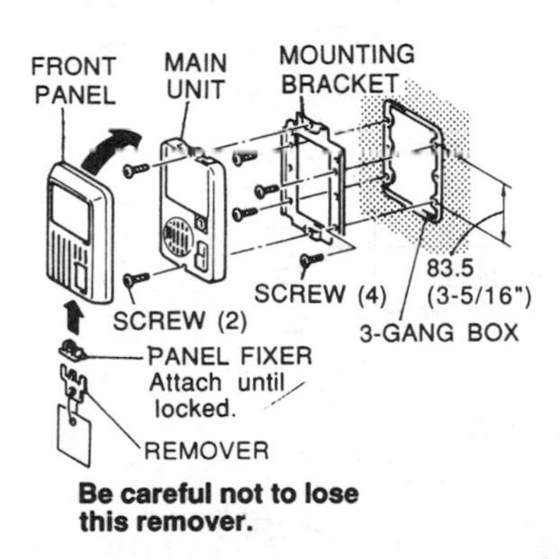

Be careful not to lose this remover.

Surface-mounting on MYW-R box;

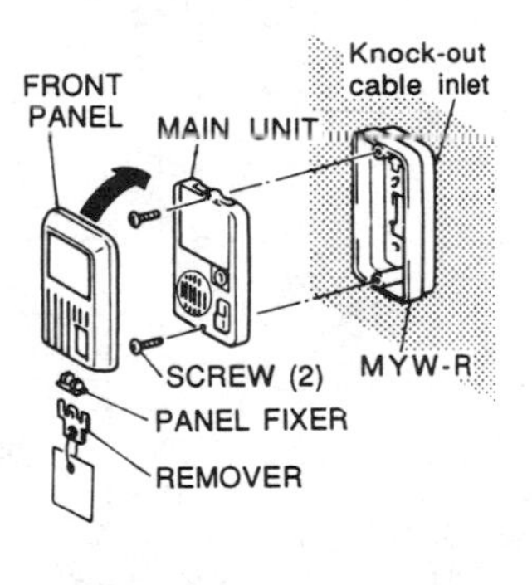

How to remove panel;

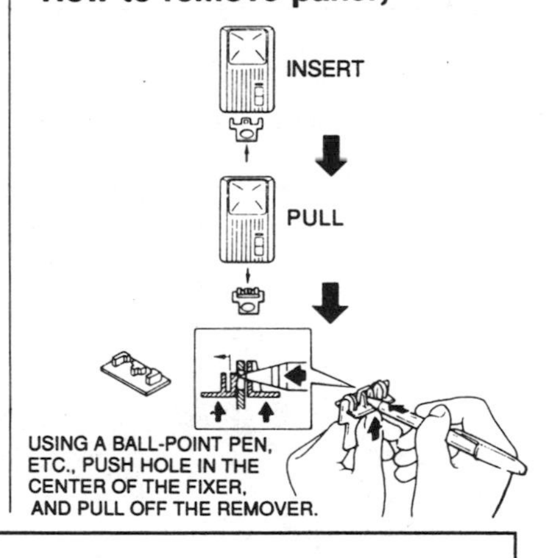

(4) Power supply connection/installation;

PS-18YC, 18YD power supply is shipped with mounting bracket. Attach the bracket to PS-18Y back chassis with two screws, and then fix the PS-18Y unit to wall with two wood screws.

For PS-18YC mounting, you have to remove two screws from the power supply unit. Attach the bracket to power supply with the supplied two screws.

DO NOT REMOVE the other two screws on PS-18YC unit.

> Keep the PS-18YC, D power supply away from the radio receiver, etc., as it will induce noise in communication.

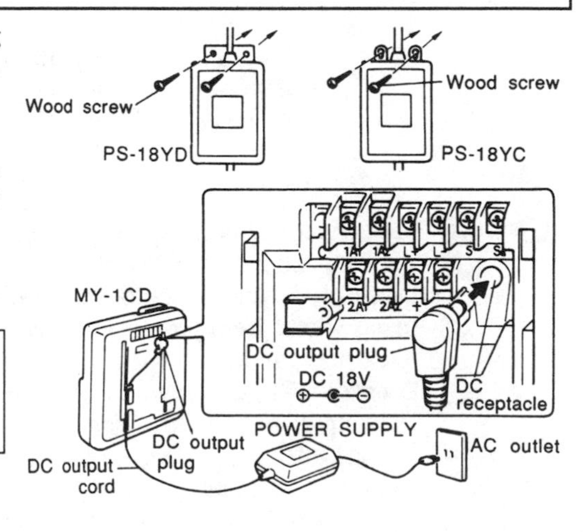

Figure 7–2. *Continued.*

SYSTEM COMPONENTS

MODELS	DESCRIPTION	REMARKS
MY-1CD	Room station with video monitor.	Surface-mounts on SINGLE-GANG BOX or wall.
MY-DC	Video door station.	Semi-flush mounts on 3-GANG BOX.
PS-18YC	Power supply. AC120V, 50/60 Hz. UL/CSA approved.	For wall-mounting.
PS-18YD	-do-, AC220-240V, 50/60Hz.	
IER-2	Call extension.	
EL-9S	Door release. AC12V, 0.35A.	Requires a bell transformer (PT-1210N for USA).

6 WIRING DIAGRAM

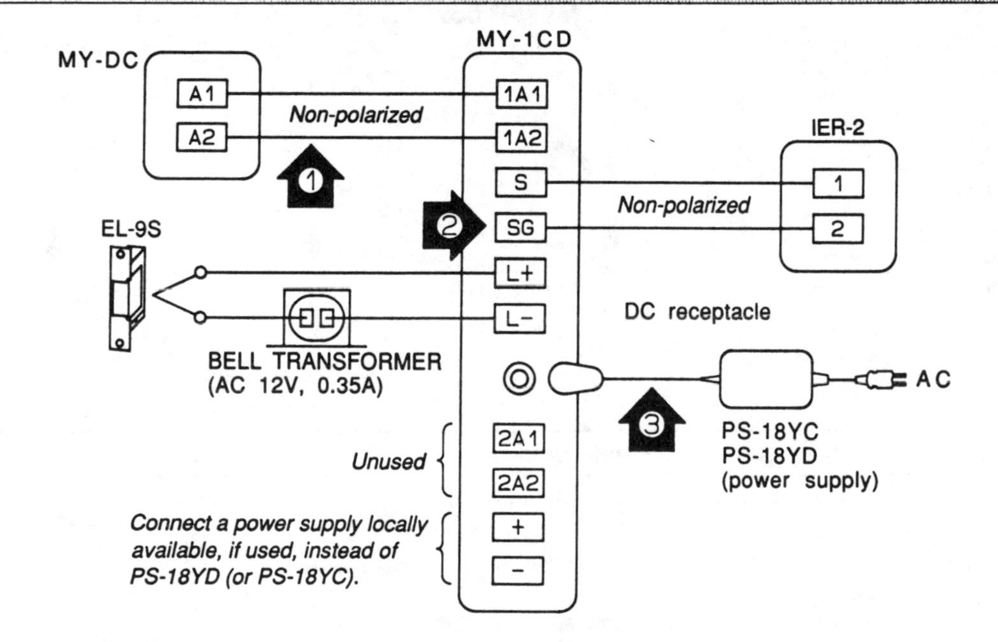

① For connecting between video door station and MY-1CD, do not use coaxial cable, or two separate and individual wires or pair in multi-conductor cable.
Use only a straight 2-conductor cable.

② In case noise from power supply lines is induced, take SG terminal on the MY-1CD unit to earth ground.

③ Locate MY-1CD unit within convenient reach of AC outlet.
DC output cord length : approx. 1.9 m (6' 3").
Do not attempt to lengthen the DC output cord. It will seriously affect the quality of the picture and communication.

Figure 7–2. *Continued.*

OPERATIONS

* Receiving a call from door station;

Momentarily depress CALL button.
4-stroke chime sounds and picture appears instantly.
Pick up handset to reply. When the picture shuts off, depress MONITOR button.
To activate the electric door release, depress and hold DOOR RELEASE button. The visitor may open the door while the buzzing sound is heard.

NOTE; *If you do not reply, the picture shuts off approximately 45 seconds after chime tone sounds.*

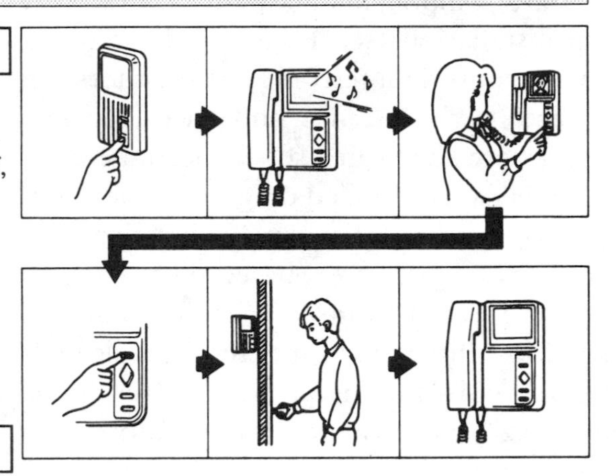

* Monitoring the entrance area;

Depress MONITOR button. The picture appears on the monitor for approximately 2 and half minutes.
Use PAN-TILT button to monitor the area widely (Camera-gearing sound is heard).
Depress MONITOR button again to cancel.

NOTE; *Inside sound will not be heard at the door station.*

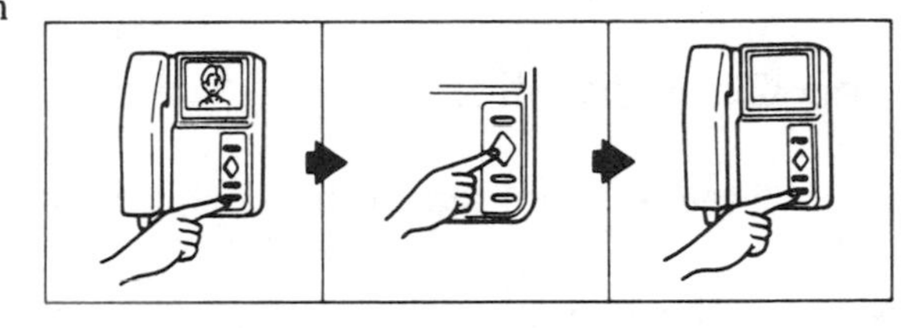

* PAN-TILT button operation;

When you can not see a caller(s) image(s) fully in the monitor, use PAN-TILT button to capture the face(s) right in the center of the monitor. Clattering sound means the camera unit no longer moves.
Release from the button.

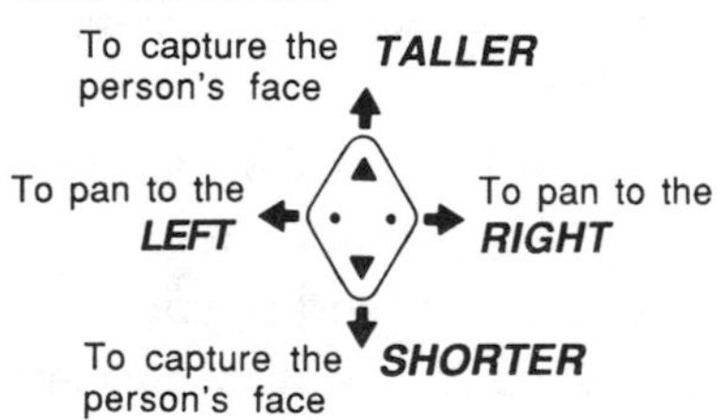

* Backlight button;

When a caller's face appears dark, press BACKLIGHT button to brighten the face.

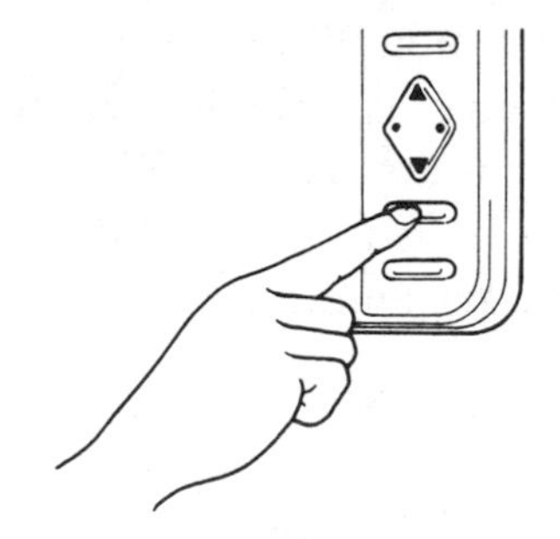

* Adjusting contrast;

LEFT DAY. NIGHT — SOFT
MIDDLE DAY. NIGHT — MEDIUM
RIGHT DAY. NIGHT — STRONG

(MY-1CD bottom, left)

At the entrance where it is dark even in the daytime, put the switch to NIGHT position and turn down the brightness.

* Adjusting brightness;

BRIGHT

DARKER BRIGHTER

(MY-1CD bottom, right)

* Adjusting chime volume;

TONE VOL.

LOWER HIGHER

Adjust the chime tone volume according to the room conditions.

Figure 7–2. *Continued.*

the camera unit. The manufacturer's suggested retail price is about $1,300.

Separate components allow more flexibility during installation (see Figure 7–2). You can mount the outside intercom where visitors can easily reach it, for instance, and place a separate camera where they can't see it. Separate cameras can be easily installed on gates, near swimming pools, and at other outside locations.

Monitors range in size (based on the diameter of the screen) from 4 inches to about 20 inches. Most integrated units have a built-in 4-inch black-and-white monitor. The 9-inch and 12-inch sizes are popular for monitors used with separate units.

Lighting Considerations

Different cameras need different amounts of light to view a scene properly. Whether it comes from the sun, from starlight, or from light bulbs, the amount of light needed is measured in "lux."

The fewer lux a camera needs, the more adaptable it is to nighttime viewing. Most color cameras need a minimum illumination of 3 lux; black-and-white cameras usually need only 1 lux. Some security units require 0 lux because their cameras have built-in infrared diodes that produce the necessary amount of light.

Buying Tips

You'll find the lowest-priced video intercoms at large hardware and builders' supplies stores. Many low-cost models are clones of expensive popular brands.

To get the best price on a brand-name model, ask the manufacturer for product literature and for the phone numbers of local distributors. Call several distributors to get their current prices. (Be sure the quotes include shipping charges.) Many distributors actively compete with one another, frequently changing their prices.

Installation Tips

Is there a "best way" to install a video intercom? The installation method will depend on the model you've chosen, where you want to use it, and how you want to use it. Most manufacturers will supply you with installation instructions, templates, and a wiring diagram. The following tips may make the job easier.

In a typical installation, you'll first need to decide where to mount the camera and the monitor units. Regardless of its minimum lux requirement, the camera should be placed where there is always enough light for you to read a book page (a porch light or streetlight may provide enough light at nighttime). With less light, you probably won't be able to see people well on your monitor screen. Don't position the camera so that it will be subjected to direct, glaring sunlight. Another consideration is that the place you choose should allow for easy wiring access to the indoor monitor.

The monitor unit can be placed in any location that's convenient for you (often, in the room where you tend to spend the highest percentage of your waking time) and near an electrical outlet. It's usually best to run your cable (you'll be using coaxial, 2-wire, or 4-wire cable) before mounting the camera and monitor units. In that way, if you have trouble getting the cable to the desired locations, you can choose other mounting locations without leaving unsightly screw holes.

RECOMMENDED RESOURCES

(See HM Sourcelist for more information.)

Aiphone Corporation—Video intercoms

Linear—Video intercoms

Novi International—Video intercoms

NuTone—Video intercoms

Profile Consumer Electronics—Video intercoms

Sony Security Systems—Cameras, CCTVs, monitors, peripherals

Ultrak—Cameras, CCTVs, intercoms, peripherals

Vicon Industries, Inc.—Cameras, CCTVs, monitors, peripherals

EXPERT ADVICE

Harold Frank

I asked CCTV expert Harold Frank of H. Frank & Associates in Palm Beach, Florida, for tips on choosing and installing CCTVs for home use. Mr. Frank is a certified protection professional who has specialized in high-tech electronic security design, procurement, training, and installation since 1965.

Q. When securing a home, in what ways do you typically use CCTVs?

A. For a home, I would install a CCTV primarily to monitor who comes through the front door or back door. If children live there, I would also install cameras in playrooms. Generally, I don't use many cameras in a home. I recently installed a system in a 5,000-square-foot home and used only four cameras.

Q. What tips would you give a homeowner who's installing a CCTV camera?

A. It's important for the camera to be positioned so that the light is between the camera and the scene being viewed. Make sure that the light doesn't come in and blind the camera. Too much direct light will result in a poor picture on the monitor.

Q. Where do you prefer to install monitors in a home?

A. I'll install them in several places in a home to make sure that they're readily accessible. The places I use most often include the master bedroom, the kitchen, and the family room.

Q. For a typical home installation, would you recommend using pan-and-tilt units?

A. A pan-and-tilt can be useful but, generally, it isn't necessary if all the homeowner wants to do is see who's coming to the door. By using the proper lens, a person can usually see a wide enough area.

Q. In what situations would you recommend that a homeowner install a covert system?

A. Most of the systems I install in private homes are overt. Sometimes I'll install a covert system because the people in the house think CCTV cameras are ugly. The only time I had a good security reason to install a covert camera was when I installed one at the top of a gigantic two-floor stairway so the family could see who was coming up the stairs.

Q. Under what circumstances do you think it's better for homeowners to hire a professional installer rather than installing their own CCTV systems?

A. People who call me to do a private home are usually business executives who fear being kidnapped. They need a state-of-the-art system that works automatically, and they're willing to pay for it. Another time someone should consider calling a professional is when the person is building a home, because the professional may be able to save a homeowner money at that point. In most cases, do-it-yourselfers would be better off installing their own systems.

8

Preventing and Surviving Home Fires

Opening

Vying with Canada during the past two decades, the United States continues to have one of the worst fire-death records among industrialized countries. Most fire-deaths in North America occur in homes and could have been avoided if the victims had properly understood the threat of fire and had taken simple precautions.

Many people in the United States and Canada don't take fire safety seriously. During school fire drills, teachers and students stand outside talking and giggling while the fire marshal reviews reporting and exit procedures. We tend to feel sympathy for a person who experiences a home fire. In Great Britain and other countries, fire victims are penalized for their carelessness. Perhaps the contrast in attitudes has something to do with the difference in fire-death rates.

This chapter looks at how home fires occur, how you can avoid them, and how you and your family can survive one. I'll tell you about some important fire safety devices and show you how they can best be used.

CAUSES AND CURES

According to the United States Fire Administration, most home fires can be traced to smoking, cooking, heating equipment, and electrical appliances.

More civilians die in fires related to in-house smoking than in any other type of fire. Over 90 percent of fire-deaths each year are the result of someone's falling asleep or passing out while a lighted cigarette was being held or was burning out on a nearby furniture surface or in a waste-basket. Mattresses, stuffed chairs, and couches often trap burning ashes for long periods of time while releasing poisonous gases. Many people are killed by the gases rather than by flames.

The best way to avoid smoking-related fires is to not smoke and not allow others to smoke in your home. If you're a smoker or allow smoking in your home, be sure that sturdy, deep ashtrays are placed in every room. Always douse butts with water before dumping them in the trash, and check under and behind cushions for smoldering butts before leaving home or going to bed. *Never* smoke when you're drowsy or while you're in bed.

The kitchen, where people work with fire most frequently, is the leading room-of-origin for home fires. Here are some simple things you can do to virtually eliminate the risk of ever having a major kitchen fire:

1. Keep your stove burners, oven, and broiler clean and free of grease;
2. When stovetop cooking is underway, never leave it unattended;
3. Turn handles of pots and pans away from the edges of the stove while cooking;
4. When you cook, wear short sleeves or keep your sleeves rolled up (to avoid dragging them near the flames);
5. Make sure no towels, paper, food wrappings or containers, or other flammable items are close to the stove;
6. Don't use towels as potholders (they ignite too easily);
7. Never store flammable liquids in the kitchen.

You may still occasionally have a small grease fire while cooking. (Be especially careful when frying foods.) Be prepared to respond immediately to such a fire. Turn off all burners of the stove and quickly cover the burning pan with a large metal lid. If no metal lid is at hand, pour a large quantity of flour onto the burning area to smother the flames while you get a cookie sheet that can be placed over the surface of the pan to seal off oxygen—or activate your fire extinguisher (if you own one). *Don't* pick up the pan and carry it to the sink. You may burn yourself, or spill burning grease, or drop the pan and start a fire on the floor.

Although more fires start in kitchens than in any other rooms, cooking isn't the main culprit. The number-one cause of home fires is heating equipment. Nearly one-fourth of home fires involve space heaters, fireplaces, or wood stoves.

To avoid a heating equipment fire:

1. Make sure any heating equipment you buy has been tested and approved by an independent testing laboratory (such as Underwriters Laboratories);
2. Be sure to follow manufacturers' instructions when installing and using the equipment;
3. Never leave flammable materials near heating equipment;
4. If you have a space heater, always keep it at least 36 inches away from anything

combustible, including wallpaper, bedding, clothing, pets, and people;

5. At the start of each heating season, make sure your heating system is in good working order. Check standing heaters for fraying or splitting wires and for overheating. If you notice any problems, have all necessary repairs done by a professional.

During a typical year in the United States, home appliance and wiring problems account for about 100,000 fires and over $760 million in property losses. Many fires could have been prevented if someone had simply noticed a frayed or cracked electrical cord and had it replaced.

You may think most of my fire safety suggestions are so obvious that they don't need to be stated. They *are* obvious, but everyday fires occur because someone failed to take one of those simple precautions. While I was preparing this book, for example, my 8-month-old nephew was severely burned in a fire at his grandmother's house. Although there's disagreement about how the fire started, the fire inspectors said it started in the kitchen, where cans of gasoline were being stored. (Fortunately, after extensive plastic surgery, my nephew is doing all right.)

SMOKE DETECTORS

A smoke detector is the single most important home safety device (see Figure 8–1). About 80 percent of all fire-deaths take place in homes not equipped with working smoke detectors. Most fatal home fires occur between midnight and 4:00 A.M., when residents are asleep. Without a smoke detector, you may not wake up during a fire, because smoke contains poisonous gases that can put you into a deeper sleep.

The vast majority of homes in the United States have at least one smoke detector installed, but most of the detectors don't work because their batteries are dead or missing. Simply having a smoke detector isn't enough. It has to remain in working order to help you stay safe.

There are two basic types of smoke detectors: ionization detectors and photoelectric detectors. They work on different principles, but either type is fine for most homes. Considering that many models sell for less than $10, it's foolish not to have several working smoke detectors in your home.

Smoke detectors should be installed on every level of a home, including the basement. A detector should be placed directly outside of each bedroom. The best location is one that is away from air vents and about 6 inches away from walls and corners.

Test your smoke detectors once a month to make sure they're in good working order. If they're battery operated, replace the batteries as needed—usually about once a year. Some models sound an audible alert when the battery is running low. Don't make the mistake of removing your smoke detectors' batteries to use them for operating something else.

FIRE EXTINGUISHERS

A fire extinguisher can give important protection if you have the right model and know how and when to use it. If you use the wrong type, you can actually make a fire spread.

There are several types of fire extinguishers, and each type is designed to extinguish fires originating from particular sources. The main types are:

Class A: wood, paper, plastic, and clothing fires;

Class B: grease or flammable liquids fires;

Figure 8–1. A well-placed, working smoke detector is the single most important home safety device. (Courtesy of BRK Electronics)

Class C: electrical equipment and wiring fires;

Class ABC: virtually all types of home fires.

The Class ABC extinguisher, or another model that combines two or more classes, is recommended for home use.

Buy a fire extinguisher that everyone in your home will find easy to use. It won't be much good if no one is strong enough to lift it. Look for a model that has a pressure gauge dial. You will then know at a glance when the pressure is low and the extinguisher needs to be refilled.

When you buy a fire extinguisher, read the instructions carefully and reread them every few months. You should be ready to use it correctly and without hesitation at any time. To use most extinguishers, you should stand at least 8 feet away from the fire, remove a pin from the extinguisher, aim the nozzle at the base of the fire, and squeeze the trigger.

Don't ever conclude that because you own a fire extinguisher you have no need for your local firefighters if a fire breaks out in your home. You are not a trained firefighter, and the visible flames are only one lethal element of a fire. Unless the whole incident is over in a matter of seconds, call your fire department. You should use your extinguisher only to put out a small fire and only if the fire is not between you and your only means of escape. Many small fires quickly spread and become uncontrollable and life-threatening. Before trying to put out any fire, make sure you have a way to escape and use it immediately if the extinguisher is not large enough to put out all of the fire.

ESCAPE LADDERS

If you live in a multiple-story home, plan a way to escape safely from windows located above the ground floor. You might install rope-ladder hooks outside each upper-floor bedroom, and keep a rope ladder in each of the bedroom closets. Another option is to use a fixed ladder, such as the Redi-Exit.

The Redi-Exit is a unique ladder that is disguised as a downspout when not being used (see Figure 8–2). Its shape discourages people from trying to use it to gain entry into a home. From an upper-floor window, you can open up the Redi-Exit by striking down on a release knob (see Figure 8–3). The unit can be installed on a new or existing home.

FIRE SPRINKLER SYSTEMS

Studies by the U.S. Fire Administration indicate that the installation of quick-response fire sprinkler systems in homes could save thousands of lives, prevent a large portion of fire-related injuries, and eliminate hundreds of millions of dollars in property losses each year.

Sprinklers are the most reliable and effective fire protection devices known, because they operate immediately and don't rely on the presence or actions of people in the building. Residential sprinklers have been used by businesses for over a century, but most homeowners haven't considered installing them because they are misinformed on sprinklers and misunderstand their use.

One misconception about residential sprinklers is that all of them will be activated at once, dousing the entire house. In reality, only the sprinkler directly over the fire will go off, because each sprinkler head is designed to react individually to the temperature in that particular room. A fire in your kitchen, for example, won't activate a sprinkler head in your bedroom.

Another misconception is that fire sprinklers are prohibitively expensive. A home sprinkler system can cost less than 1 percent of the cost of a new home—about $1.50 per square foot. The additional cost may be minimal when spread over the life of a mortgage. You may find that a home sprinkler system virtually pays for itself in homeowner's insurance savings. Some insurers give up to 15 percent premium discounts for homes with sprinkler systems.

If you can't see your way clear to installing a full home sprinkler system, consider one that protects one of your most vulnerable areas—your kitchen stove. The Guardian is the first automatic range-top fire extinguisher available for home use (see Figure 8–4). It was developed for U.S. military use after a 1984 study identified cooking-grease fires as the number-one cause of fire damage and injuries in military-base housing. The patented system uses specially calibrated heat detectors to trigger the release of a fire-extinguishing chemical (see Figure 8–5).

When the chemical is released, the system automatically shuts off the stove. In laboratory tests, The Guardian has been found to detect

Figure 8–2. When not in use, the Redi-Exit looks like a downspout. (Courtesy of Karsulyn Corporation)

Figure 8–3. After being released, the Redi-Exit becomes a ladder. (Courtesy of Karsulyn Corporation)

Figure 8–4. "The Guardian" is a range-top fire extinguisher designed for quickly putting out cooking-related fires. (Courtesy of Twenty First Century International Fire Equipment & Services Corporation)

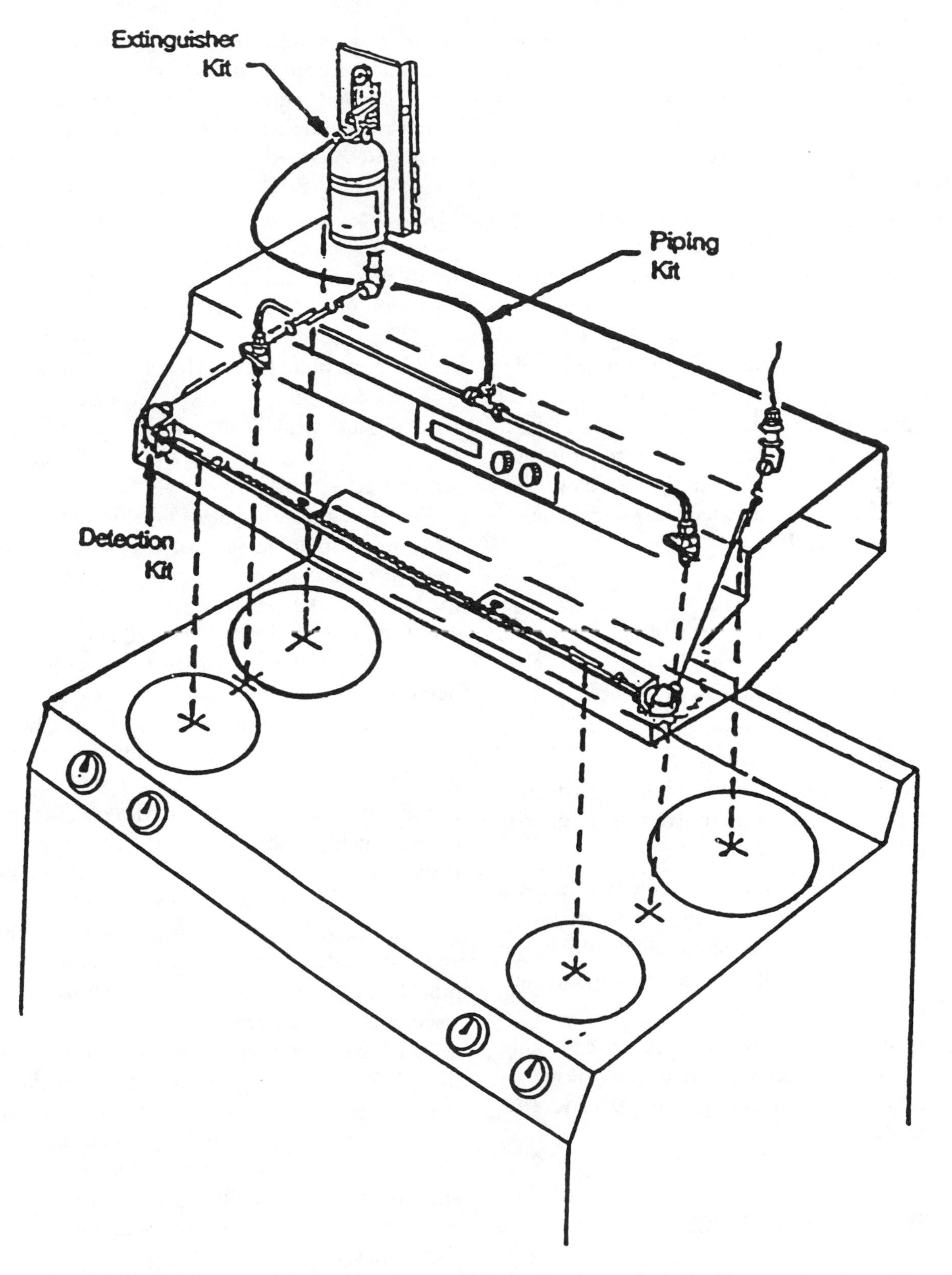

Figure 8–5. "The Guardian" is a patented system that can be installed in a range-top. (Courtesy of Twenty First Century International Fire Equipment & Services Corporation)

and extinguish stove-top fires within seconds—but not to activate under normal cooking conditions. You can have it installed so that it also activates an alarm inside your home.

The system is UL-listed and sells for under $1,000.

The Guardian combines a fire detection assembly, an extinguisher assembly, and a chemical distribution system into a single automatic unit. The fire detection system can be neatly installed under any standard range-top hood. Cables connect it to the extinguisher assembly, which is housed in the cabinetry above the stove top. A pressurized container stores a fire-extinguishing liquid, which is distributed through stainless steel piping to the under-hood nozzles.

Here's how The Guardian responds when a stove-top fire starts:

1. Extreme heat from the stove-top fire causes any of four fusible links in the under-hood detection assembly to separate, releasing tension on a cable.
2. When the cable tension is released, a tension spring automatically opens the extinguisher valve, discharging the liquid extinguishing mixture through the piping.
3. The mixture flows through two nozzles positioned directly above the stove-top burners, and a controlled discharge smothers the fire and guards against another fire's starting.
4. While the extinguishing mixture is being released, a microswitch activates a switch that shuts off the gas or electric fuel source.

SURVIVING A HOME FIRE

To ensure that you and your family will be able to get out alive during a fire, you need to plan ahead. All members of the household, including small children, should help develop an escape plan and regularly practice using it.

It isn't enough to just say what you plan to do in case of a fire; you may have only seconds to get out, and the smoke may be so thick and black that you won't be able to see where you're going. Only through practice will you be able to react quickly and do almost routinely what you need to do to survive.

Make sure all potential escape routes are readily accessible. Check that your windows are not painted shut; remove furniture that is blocking a door; adjust locks that are too high for children to reach; and so on. Take care of any obstacles *right away*.

Establish a meeting place outside and not too close to your home (a spot near a designated tree, or on a neighbor's porch). Agree that all members of the household will go there and wait together for the fire department. Everyone should know how to call for help—either at a neighbor's home or by using a fire box.

No one should go back into a burning home, even if someone is still unaccounted for. If you go back into your home, you're endangering not only yourself but anyone who's inside. Fire grows quickly, and it rushes to wherever there is oxygen. As you open windows or doors in a burning home, you're making the smoke and flames spread faster. It's better to stay outside and wait for the firefighters. They'll arrive very quickly and will have the equipment and skills to rescue anyone left inside.

Here are some key actions that everyone should remember. If you encounter smoke on your way out, try to use an alternate exit. If you must escape through smoke, crawl along the floor, under the smoke, where the air is cooler and cleaner. If your clothing catches on fire, stop, drop to the ground, and roll to extinguish the flames.

If you're in a bedroom and you hear a smoke detector, leave quickly through a bedroom window if possible. If the room is too high off the ground, or if you can't safely get out of the window, feel the door from the bottom up to find out whether it's warm. Don't touch the door knob; it may be hot. Don't open the door if it feels warm. If the door is cool, place your shoulder against it and open it slowly. If you don't see flames and an exit is near, quickly crawl to safety. Once you are out of the building, call the fire department immediately. Don't go back into the building *for any reason*.

If the bedroom door is hot and you can't safely climb out of a window, stuff rags or rolled-up clothes under and around the door and in every other gap or opening that may allow smoke to enter the room. If you can safely open the window, hang a rag or piece of clothing out of it. That will let the firefighters know where you are.

WHAT TO DO AFTER A FIRE

If you have a home fire, take these actions as quickly as you can afterward:

1. Immediately call your insurance company or the insurer's agent, and then call your mortgage company.
2. Don't let anyone into your home without first seeing proper identification. Criminals may try to take advantage of your vulnerable situation.
3. Make sure all your utilities are turned off. If you're in a cold climate and you expect your house will be empty for a long time, drain the water lines.
4. If possible, board up all holes in the roof, doors, windows, and other entry points.
5. Protect all undamaged property, to avoid further damage.
6. Don't clean up until after your insurance company has inspected the damage.
7. Make a list of all your damaged property. If possible, include the model numbers, serial numbers, dates and places of purchase, and purchase prices. The more details you have about your property, the

After a Fire: Some Clean-Up Tips

1. Clothing. To remove smoke odor and soot from clothing that can be bleached, try a mixture of 1 gallon of warm water, 4 to 6 tablespoons of trisodium phosphate ("TSP," sold in hardware stores), and 1 cup of Lysol or any household chlorine bleach. After mixing well, add it to the clothes you're washing, and rinse them with clear water. Then dry them thoroughly.

   To remove mildew from clothes, use soap and water to wash the fresh stain. After rinsing, dry them in the sun. If the mildew stain is still present, rub it with lemon juice and salt, or a diluted solution of household chlorine bleach. (Note: Always test colored garments before using any treatment.)

2. Cooking items. Wash all your pots, pans, and cooking utensils in hot soapy. water. After drying them, use a fine-powdered cleaner to polish them. Copper and brass can be polished with salt

sprinkled on a piece of lemon or salt sprinkled on a cloth saturated with vinegar.

3. Food. Wash your canned goods (and food in jars) in detergent and water. If the labels come off during cleaning, mark the contents on the can or jar with a grease pencil. (Note: Don't use canned goods if the cans are dented, rusted, or bulged.)

4. Refrigerators and freezers. To remove odor from a refrigerator or freezer that has stopped running, wash the inside walls with a solution of baking soda and water, or a mixture of 1 cup of vinegar or household ammonia and 1 gallon of water. After cleaning the refrigerator or freezer, place an open container of baking soda or a piece of charcoal in it to absorb odors.

5. Rugs and linoleum. When water gets underneath linoleum, it can cause odors and warp a wood floor. If that happens, remove the entire sheet of linoleum. If it's brittle, use a heat lamp to soften it so that it can be rolled up without breaking. By removing the linoleum carefully, you may be able to re-cement it after the floor dries completely.

   Throw rugs can be cleaned by beating, sweeping, or vacuuming, and then shampooing. They should be dried immediately by laying them flat and exposing them to a circulation of warm, dry air. You can use a fan to speed the drying process.

6. Mattresses. If the inner spring of your mattress needs reconditioning, take it to a company that builds or repairs mattresses. If you need to use the mattress temporarily, dry it out in the sun and cover it with a rubber or plastic sheet.

7. Floors and furniture. Soot and smoke can be removed from floors and furniture by cleaning them with the same mixture as for your clothing: 1 gallon of warm water, 4 to 6 tablespoons of TSP, and 1 cup of Lysol or any household chlorine bleach. Wear rubber gloves when cleaning. After washing each area, rinse it with clear water and dry it thoroughly. Don't dry wood furniture or fixtures in the sun (the wood may warp).

8. Walls. Use a mild, paint-safe soap or detergent to wash your walls. Working from the top down, clean a small area at a time and then immediately rinse the area with clear water. Wash ceilings first. Don't repaint the walls or ceiling until they have completely dried.

9. Locks and hinges. Thoroughly clean and oil all of the hinges on your doors and windows. Contact a locksmith to clean and oil the internal parts of your locks. The servicing will help to ensure that your hinges and locks will continue to work properly.

better off you'll be when dealing with your insurance company.

8. If your home is too damaged for you to live in and you need temporary shelter, call your insurance company, the local Red Cross chapter, or the Salvation Army for help. Other possible sources of help include local churches and synagogues, and civic groups such as the Lions Clubs International and the Rotarians.
9. Keep all receipts for additional living expenses and loss-of-use claims.

Be wary of uninvited insurance adjusters who may contact you after hearing a report of the fire. If you have a complete inventory of your property and it's readily accessible, an insurance adjuster probably can't do any more for you than you can do yourself.

RECOMMENDED RESOURCES

(See HM Sourcelist for more information.)

BRK Electronics—Fire extinguishers, smoke detectors

Generation Two—Smoke detectors

International Association of Home Safety and Security Professionals (IAHSSP)—Trade association

Karsulyn Corporation—Escape equipment

National Fire Information Council—Trade association

National Fire Protection Association—Trade association

National Fire Sprinkler Association, Inc.—Trade association

Twenty First Century International Fire Equipment & Services Corporation—Fire extinguishers

United States Fire Administration—Federal agency

9

Keeping Thieves Out of Your Car

Car theft is the nation's fastest growing major crime, according to the U.S. Department of Justice. In 1992, one of every 42 automobiles was stolen or had its contents or accessories taken.

Some vehicles are stolen by juveniles for joy riding, but the greatest threat to your car is the professional car thieves who search for cars to sell or strip for parts. Using special tools and methods, within minutes they can take whatever model they need.

Expensive and late-model cars aren't the only ones the professional thieves want. The average value of a vehicle at the time of theft is only about $5,000. When sold through a "chop shop," some car parts (engines, doors, hoods, fenders, and so on) can be more valuable than an intact car.

There are many precautions you can take, and several low-cost devices are available to keep thieves out of your car. Some highly advertised anti-theft devices only make car thieves laugh. This chapter discusses which devices work, which don't, and why.

ANTI-THEFT DEVICES

The three basic types of anti-theft devices for vehicles are cutoff switches, supplemental locks, and alarms. Cutoff switches disable a car that someone has tried to start without using the right key. Some cutoff switches allow a car to run for a short while; others prevent the engine from turning over. Those that allow the car to run can be dangerous not only to the unauthorized driver but also to others on the road.

Supplemental locks are used for securing various parts of a car—doors, trunks, and steering wheels. The most popular type is the steering-wheel lock, which is sold under many brand names. One style is a cane-shaped bar that locks the steering wheel to the brake pedal (see Figure 9–1). Another style uses the dashboard as an obstruction to prevent the steering wheel from turning (see Figure 9–2). Most steering-wheel locks are easy to saw off, pick-open, or bypass.

When a steering-wheel lock is used properly, however, it lets thieves know that the car is owned by a security-conscious person, and may make them wonder what other security measures have been taken. Any car thief would prefer to "hit" a car that doesn't have a steering-wheel lock in place. Unfortunately, thieves are more likely to find a steering-wheel lock in a car's trunk than on its steering wheel, because many owners find the constant need to put the lock on and take it off too much of a

Figure 9–1. Many steering-wheel locks connect the steering wheel to the brake pedal. (Courtesy of Kryptonite Corp.)

Figure 9–2. The "Thundarbar" is a combined steering-wheel lock and self-contained alarm. (Courtesy of Carbrella Motoring Accessories)

nuisance. Steering-wheel locks sell for between $40 and $125 in auto supply stores, department stores, hardware stores, home centers, and locksmith shops.

CAR ALARMS

Some alarms can be useful for deterring and catching car thieves, but many models are virtually worthless.

There are two basic types of car alarms. A *passive alarm* automatically arms itself, usually when the car is turned off and the last car door is closed. An *active alarm* requires you to do something special—push buttons or flip a switch—to arm the car. Because it's easy to forget to arm a system, the passive systems provide more reliable security.

Many insurance companies give owners of a passive system a 5 percent to 20 percent discount off the comprehensive portion of their car insurance. The states that require insurance companies to do so are: Illinois, Kentucky, Massachusetts, Michigan, New York, and Rhode Island.

Here's how a typical car alarm works. The alarm consists of detection devices, a siren or other warning device, and a control module (see Figure 9–3). All the components are connected to one another by wire or radio waves to form a complete electrical circuit. The detection devices are strategically placed so that a thief cannot take the car without causing a break in the circuit. When a door is opened, for instance, a circuit break should occur, causing the sirens, lights, and other warning devices to activate.

Most car alarms have current-detection devices; they sense the current drop that occurs when a car's courtesy light comes on. That feature alone usually isn't enough to protect a vehicle, because a thief can gain entry in ways that don't activate the courtesy light. Other types of detection devices respond to the sound of breaking glass, to vibrations, or to motion within a car. Generally, the more types of detection devices a system uses, the more effective it will be.

The most common detection device is the pin switch, which senses the opening of a door or trunk. The thin, spring-loaded switch is installed in a small hole in the frame around the door (or trunk) so that it is compressed when the door is closed. While it's compressed, the switch helps to complete an electrical circuit. When the door is opened, the switch springs up, breaking the circuit. The switch has a

Figure 9–3. A typical car alarm consists of a siren, a control panel, and several detection devices. (Courtesy of the David Levy Company, Inc.)

single wire; the car's metal chassis is the ground portion of the circuit.

A motion detector can be used to protect against thieves who use glass cutters to gain entry, because it can sense cut glass falling into the car. It can be especially useful for vehicles that have vent windows or sliding windows.

Some alarm systems use strobe lights or a car's headlights or parking lights as warning devices. A light emitting diode (or LED) installed on a door or dashboard can also be used as a warning device. It continuously flashes when the alarm is armed, letting thieves know the car is protected. It flashes at a faster rate whenever the alarm has been activated, letting owners know the car has been tampered with.

The most popular type of warning device is a siren. Many models sound at 100 or 113 decibels (db); better models sound at a minimum of 120 decibels. Some locales have restrictions on the types of sirens that may be used. You should use one that automatically shuts off and resets itself after a certain period of time. A siren that blasts continuously will drain your battery and be a nuisance to people nearby.

The control module of a car alarm is the brain of the system. When it learns from a detection device that the electrical circuit has been broken, the control module decides when to activate the sirens and other warning devices. It usually has an entry/exit delay feature to allow you enough time to get into and out of your car without triggering the alarm. Depending on the model, you can adjust the delay time within 12 to 40 seconds. The shorter the delay, the less time a thief will have to disable the system before the alarm is triggered. No delay should be set for the trunk and hood.

The biggest problem with most car alarms is that they depend on passersby to intervene. Car thieves know that if they can disarm an alarm quickly, they will attract little or no attention from passersby. Even if someone does challenge the car thief, all he has to say is that it's his car. Few people care enough about strangers' cars to risk offending someone who may or may not be a thief.

Rather than relying on passersby to tell a thief to leave your car alone or to call police, you can carry a remote pager in your pocket. It will warn you whenever your car alarm is activated. Some models allow remote arming and disarming of the alarm system.

The advertised range of a pager is often the maximum range that can be achieved under the best conditions. The actual range you need will depend on where your car is and how far away from it you'll be. If your car is in an underground parking garage and you're in a high-rise building, for instance, the pager will have less range than if you and the car were in the same open field. The more obstructions between you and the vehicle, the less range the pager will have. Good pagers have a range of over 2 miles in most situations.

Special Features in Car Alarms

Car alarms are no longer designed only for security. Many offer a variety of conveniences. Some car alarms allow you to use a remote control to: start or shut down your car's engine, heater, and air conditioner; lock and unlock the doors; open the trunk; control windows and a sunroof; activate the interior courtesy light; flash the parking lights; arm and disarm the alarm; monitor and recharge the car's battery; and control electric garage doors.

Three special security features to look for when choosing an alarm are: a self-recharging backup battery, a hood lock, and anti-scanning circuitry. A backup battery will replace your primary battery if it runs down or is disconnected by a thief. A hood lock connected to your alarm will activate the system if a thief tries to get under the hood. (The factory-installed hood locks on most cars are easy to defeat.)

Before buying an alarm that allows remote control operation, make sure it has anti-scanning circuitry. Hand-held frequency scanners can be made from parts found in any electronic hobby shop or bought by mail order. Car thieves use these scanners for rapid transmission of different codes until one is found that will disarm a car's alarm. Alarm systems with anti-scanning circuitry can detect the use of a scanner and resist being controlled by it.

Warranties

Most car alarms come with a warranty. Don't take the warranty for granted: you may need service or a replacement. All car alarm manufacturers occasionally make defective products.

Carefully read the warranty to be sure of what it does and doesn't cover. A typical warranty from a manufacturer covers parts and labor for one year. It usually applies only to defective products sold to the original retail purchaser and doesn't cover problems caused by a faulty installation.

Because rights covered by warranties vary from state to state, you may have more rights than the manufacturer has listed. For example, although most warranties say they exclude or limit "incidental or consequential damages," some states don't allow such an exclusion or limitation.

The dealer who sells an alarm to you, or the company that installs it, may also offer a warranty. If you're not certain about your rights under a warranty, consult with your attorney.

Factory versus Custom Installation

Twenty-six car lines have alarms as standard equipment. Many others provide them as options. The big problem with such alarms is that they're standardized. A thief needs only one owner's manual to learn how to disable all factory-installed alarms in a certain make or model of car.

A custom-installed system has the element of surprise. The thief won't know what type is installed, what special security features are included, or where the various components have been placed. Many custom systems cost less and are more sophisticated than factory-installed alarms.

As with home alarms, don't use a window sticker that identifies the brand or model of your system. Alarm makers like the free advertising, but the more that is known about your system, the easier it will be to defeat. Buy a generic window sticker or use one from a different alarm system.

Choosing an Installer

Will you need a professional installer? That depends on the system you choose and your ability to wire. The more components and wires a system has, the harder it is to install. The easiest alarms to install are portable, self-contained models that have a built-in current sensor, a motion detector, and a siren (see Figure 9–4).

Most alarms are placed on a car's dashboard and installed by plugging a cord into the cigarette lighter or connecting a single wire to a fuse box. A thief can disable such a unit within seconds just by unplugging the cord and removing the backup battery. Better portable models are installed under a car's hood and connected to the car's battery by a single wire.

Some manufacturers of high-tech systems provide installation instructions only to factory-authorized dealers and won't warrant a product installed by anyone else. With most systems, however, you can choose your own installer.

Car alarms are installed by burglar alarm companies, locksmiths, auto repairs shops, and audio equipment service centers. Anyone who can wire a stereo system in a car can wire most

Figure 9–4. A portable self-contained alarm is easy to install and use. (Courtesy of Quorum International, Inc.)

car alarms, but proper installation of a car alarm requires security considerations that don't apply to stereo equipment. The alarm components should be placed and wired in ways that will confuse sophisticated car thieves. If you want your system professionally installed, look for a company that specializes in car alarms or other security systems.

Find out how long the company has been installing car alarms. Ask for names and phone numbers of previous customers, and call those references to find out what problems they may have had with their systems. If any of them are unhappy with the installer, look for another installer. Their unfavorable opinion indicates that the installer didn't obtain their permission to be contacted by you—a tell-tale sign of an installer who has little respect for security.

Call your local Better Business Bureau or your state's Attorney General's Office and ask whether complaints have been filed against the company. Find out how long the installer has been doing business in your area.

After finding two or three acceptable installers, compare their prices and warranties. It's not unusual to find an installer that charges up to twice as much as others in the same area.

STOLEN-VEHICLE RETRIEVAL SYSTEMS

The latest advances in automobile security are stolen-vehicle retrieval systems. The recovery rate for stolen vehicles equipped with retrieval systems is about 95 percent, compared to only 64 percent for vehicles not equipped with them.

To use a stolen-vehicle retrieval system, a small transceiver must first be installed in a hidden recess of the car. If the vehicle is stolen, the transceiver emits a signal that is picked up by a tracking computer. The signal allows the manufacturer or the police to locate stolen vehicles quickly.

LoJack Corporation, headquartered in Dedham, Massachusetts, is the only company that supplies mobile tracking computers to police departments (see Figure 9–5). Other companies do their own tracking and charge a monthly monitoring fee for the service.

With a LoJack system, each transceiver is given a unique five-digit code that is paired with a vehicle identification number (VIN) in a State Police crime computer. When you have installed a LoJack and report a theft, police enter your car's VIN into the computer network of the State Police Broadcast System, which activates the LoJack transceiver in your car. Police cars equipped with LoJack tracking computers receive the broadcast signal of the transceiver and follow a homing procedure that takes them to the stolen vehicle—if it is still in the tracking area. You must report your car stolen before it is driven out of the range of the tracking computers.

The LoJack system is available in California, Florida, Georgia, Illinois, Massachusetts, Michigan, New Jersey, and Virginia. The basic model sells for about $600. LoJack System II comes with a starter disabler and sells for about $700. LoJack System III, at about $800, includes a starter disabler and passive alarm system.

COMBATING CARJACKERS

During the summer of 1991, in Cleveland, Ohio, a driver was waiting at a traffic light when four young men approached his car. After forcing him out at gunpoint, they jumped in and drove off.

At about the same time, a woman in Hawthorne, California, was parking her car after returning from her lunch break. Car thieves shot and killed her and stole her car.

These are two examples of a crime called "carjacking." The term was created in 1991 to identify a sudden rash of similar crimes

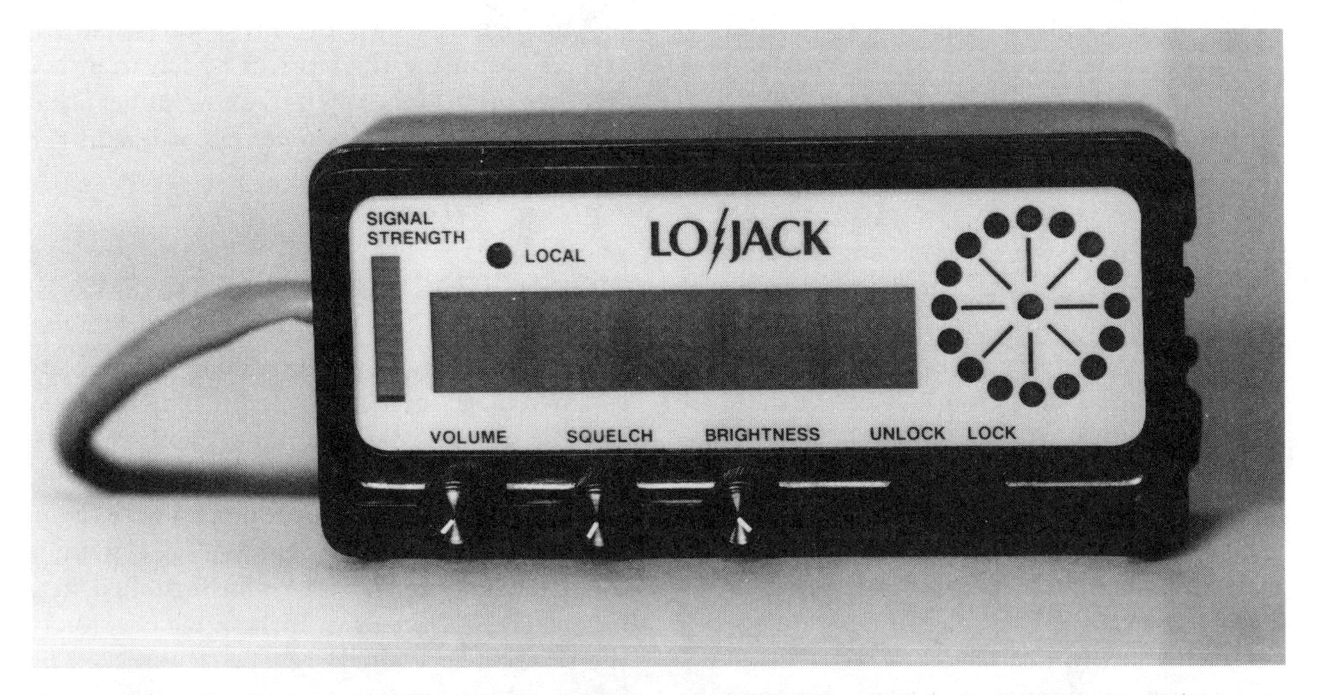

Figure 9–5. The LoJack tracking computer is installed in police cars to help them find stolen vehicles. (Courtesy of LoJack Corporation)

No-Cost Ways to Protect Your Car

Some devices can be helpful, but your own precautions can do more to protect your car than any add-ons. Here are the most important things you can do to keep thieves out of your car:

1. Whenever possible, park in a garage or an attended parking lot. Otherwise, park in a highly visible, well-lighted area between other cars. (Cars parked between others are harder to steal than those parked on the ends of a line.)
2. When you have parked your car, set the emergency brake and leave the front wheels turned sharply toward the curb. That wheel position makes it hard for a thief to tow your car away.
3. If you have to leave your car key with a parking lot attendant, get his or her name and make a show of writing it down. (Some parking attendants duplicate keys and sell them to car thieves and burglars.) Leave only the ignition key—not the keys to your doors or trunk. In that way, you will put the attendant on notice that you're security-conscious and don't mind being a pain in the butt.
4. Keep your doors locked and your windows tightly rolled up whenever you leave your car. Don't forget to take your keys with you. (One in five stolen cars were left unlocked or had the keys in them.)
5. Don't try to hide a car key in or on your car (not even with a little metal box). It takes only a few seconds for a thief to find your favorite hiding place.
6. Have your car's vehicle identification number (VIN) permanently etched into parts of your car with acid. Then place a sticker on your window to let thieves know what you've done. Because the VIN can be used to trace parts back to you, the etching will make your car's parts less valuable to thieves. Some police departments will do the etching for you and provide you with stickers at no cost.

occurring throughout the United States. Carjacking is a combination of car theft, assault, armed robbery—and sometimes murder. Carjacking happens in parking lots, at drive-in restaurants, at stop lights, at gas stations, and on highways.

Police records show that 416 carjackings occurred in Houston during the first eight months of 1991—an average of 52 a month. That figure was up 26 percent from the same period in 1990. In 1991, Dallas reported nearly 80 carjackings a month. In that year, Detroit faced nearly 300 carjackings a month. The crime was so rampant that the police formed a special carjacking task force. Several owners of downtown restaurants were forced into bankruptcy because many of their former customers became afraid to drive downtown.

Other cities that have been touched by the carjacking epidemic include New York, Chicago, St. Louis, New Orleans, Los Angeles, San Diego, San Francisco, and Portland.

There are many theories about why carjackings have suddenly increased. Some security professionals believe it's a response to sophisticated vehicle anti-theft devices: car thieves who aren't smart enough to defeat the protection systems turn to carjacking.

The flaws in that theory are: many carjackings involve cars that have no security systems, and many of the stolen vehicles aren't sold or stripped.

Carjackers seem to have little in common with people who steal cars for money. Often, a carjacker is a wild-eyed hoodlum on the run and in need of a getaway car. Drug tests indicate that as many as 90 percent of carjackers may be high on drugs. Because carjackers are often irrational and impulsive, they work in many different and unpredictable ways.

They work alone and in groups. They approach males and females of all ages. They target all types of cars at all times of day and night. It's hard to describe a "typical" carjacking.

Nevertheless, there are some common factors. By being aware of them, you'll be better able to protect yourself. Consider the following carjacking incidents:

- Farmington Hills (Michigan). While driving, a man was bumped from behind. He stopped to survey the damage. The other driver approached with a gun, ordered him out, and stole his car—leaving behind a car that had been stolen earlier.
- Houston. While shopping in an auto parts store, a 22-year-old man was shot by an armed robber. The robber fled in the man's 1990 Pontiac.
- Philadelphia. While at a gas station, a couple lost their BMW to a man who threatened them with a gun in a rolled-up newspaper.
- Portland. A County Circuit Court Judge had his $30,000 sports car stolen at gunpoint while he was in a supermarket parking lot.

At first, those examples may seem very different from one another, but they, like most carjackings, have two things in common. The perpetrators are most often young males (often gang members or drug addicts), and the victims are caught by surprise.

Carjackers like to target a person who seems vulnerable—someone who's daydreaming or someone who acts friendly or trusting toward strangers. Aside from keeping your car doors locked and your windows rolled up, the most important thing you can do to protect yourself is to stay alert—not just when you're driving or sitting at a stop light, but whenever you're walking to or from your car.

When walking to your car, have your keys ready so that you can open the door quickly (instead of having to fumble around). Be aware of anyone who may be approaching you. Before getting into your car, be sure no one is crouched in the back.

When you're driving, don't automatically stop if another car bumps yours. First notice who's in the other car. Ask yourself, "Do they look like a harmless family, or like young toughs?" If they seem suspicious, turn on your emergency flashers and signal the driver to follow you. Then drive slowly to a police station. If the people in the car seem to get upset, drive faster.

RECOMMENDED RESOURCES

(See HM Sourcelist for more information.)

Alpine Electronics of America—Car alarms

ASECO American Auto Security—Car alarms

Carbrella Motoring Accessories—Steering-wheel locks

Clifford Electronics, Inc.—Car alarms, stolen-vehicle retrieval systems

Code Alarm—Car alarms, stolen-vehicle retrieval systems

Crimestopper Security Products, Inc.—Car alarms

David Levy Company—Car alarms

DesignTech International—Car alarms

Directed Electronics, Inc.—Car alarms

Excalibur of America—Car alarms

Harrison Electronics Systems, Inc.—Car alarms

Kenwood—Car alarms, stolen-vehicle retrieval systems

Kryptonite Corp.—Steering-wheel locks

LoJack Corporation—Stolen-vehicle retrieval systems

MaxiGuard—Car alarms

Seco-Larm—Car alarms

Techne Electronics Ltd.—Car alarms

Vehicle Security Electronics, Inc.—Car alarms

Winner International—Steering-wheel locks

10

Personal Safety at Home and Away

When many people think of self-defense, they think of fighting. But fighting is a very small part of self-defense. There's also a psychological part and, when properly used, it will allow you to avoid physical battles. This chapter shows you ways to protect yourself by using psychological self-defense as well as physical self-defense.

The essence of psychological self-defense is to appear confident and alert and to be constantly aware of potential dangers. This mind-set isn't paranoia; it's just good sense. Imagine two scenarios in which a woman is walking down a street at night. First, picture the woman walking slowly and looking around nervously, as though she's lost. Now, picture the same woman walking briskly, with her head held high, and looking as though she knows exactly where she's going. That's psychological self-defense. It works because most criminals prefer to attack a person who seems helpless rather than someone who might put up a fight.

When you're using psychological self-defense, you'll try to stay away from situations that might make it hard for you to resist an attacker. You won't walk down a dark alley, for instance. At night, you'll park your car in a well-lighted area. You'll automatically check the back seat of your car before getting in.

I've studied Kung-fu, but I've never had to fight off an attacker; I've always been able to either avoid or talk my way out of potential physical attacks. Nevertheless, I think anyone can benefit from attending physical self-defense classes or studying martial arts. In addition to instilling confidence, the training can help to prepare you for the unlikely time when you need to fight to defend yourself.

PHYSICAL SELF-DEFENSE TRAINING

With proper training, any person—male or female, of any size—can defend against any other person. Even elderly persons and people with physical disabilities can learn to fight back effectively by learning about the many ways in which every assailant is vulnerable. The essence of physical self-defense is knowing how to exploit those weaknesses quickly and decisively.

Physical self-defense isn't like a sport, however. You don't necessarily have to stand toe-to-toe and brawl with someone to defend yourself against that person. If you've never studied physical self-defense, you might be surprised at how many ways there are to fight.

To learn physical self-defense, you need to train with a qualified instructor. A book or videotape can give you some tips, but only an instructor can make sure that you're maintaining the proper body mechanics and applying the techniques properly. Many fighting techniques work only when done with extreme precision.

Self-defense classes are designed to teach you some practical techniques within a short period of time. You'll learn the right way to poke attackers in the eyes or kick them in the knees. Some classes show you how to use mace or similar deterrents.

Make sure that the class you sign up for is being taught by a qualified instructor and that the training includes actually hitting another person. You can't learn to fight a person by just hitting a bag and kicking in the air. The best self-defense classes employ a heavily padded person who pretends to attack the students.

Martial arts training is much more involved. You don't learn a lot of practical applications right away. Usually, you train only 2 or 3 days a week, and most of the time during the first few months you're stretching, learning foreign words, and practicing simple movements.

Not all styles of martial arts are designed for self-defense. Some styles are used primarily for tournament fighting, which has little relationship to street fighting. In a tournament, people are heavily padded and must follow strict rules about how and where to hit.

For self-defense, the styles that require the least amount of time to learn include jujitsu, judo, and the martial arts as practiced in Indonesia, the Philippines, and Thailand—the styles that look crude and ugly. A dedicated student can use those styles effectively within 6 months of training. Styles that look graceful and flowery—like aikido and kung-fu—can also be effective for self-defense, but they usually take several years to learn.

RAPE PREVENTION

No one knows how many rapes occur each year, because the crime often goes unreported. Crime records show that every woman is a potential rape victim. Rape victims come from every social and economic background; there is no "typical" rape. The crime is committed by friends, strangers, dates, lovers, and spouses.

Rapes can occur at home, on the street, in parking lots, and at any time of day or night.

It's important to understand that rape isn't the same as a normal sex act; it's a crime of violence perpetrated by people who want to hurt, dominate, and humiliate other persons. Regardless of how a woman dresses, her appearance can't make someone become a rapist—nor will her attire discourage a rapist.

Rape is never the victims' fault, but there are precautions a woman can take that will reduce the risk of being attacked. She can use the information given throughout this book to make her home more secure, for example. She can constantly practice psychological self-defense.

Most rapes occur when a woman is isolated with a man and trapped. Women should be concerned about being placed in such a situation. When dating someone who is not known well, it's a good idea to be around lots of people. During a date, a woman needs to stay aware of her date's behavior. If he becomes hostile or makes her feel uncomfortable, it may be best for her to leave—alone.

With respect to rape prevention, psychological self-defense is mostly a matter of being wary of strangers, staying away from unfamiliar places, and trying to recognize a rapist's intentions as soon as possible. If a woman exercises her right to date, travel, and meet new people, it isn't possible to eliminate all risk of crossing the path of a rapist. That's why physical self-defense training is important.

If you experience a sexual assault, the best thing you can do is call the nearest rape crisis center. Trained volunteers and professionals are there to help you through the situation, whether or not you plan to call the police.

SELF-DEFENSE WHILE TRAVELING

Whenever you're in public, someone may be watching you and planning to attack you or take your property. Because you never know when you're being considered, you should always appear confident and alert.

When you're traveling, you're more vulnerable than usual and you need to take special

Before Going on Vacation

A home is more vulnerable when no one is home. Take these important precautions before you leave:

- Don't publicize your plans to be away. (Usually, the fewer people who know you're away, the better.)
- Set a timer to turn your inside and outside light on automatically at various times throughout the night.
- Close and lock all your windows and doors. (Don't forget the garage doors.)
- Unplug all televisions, stereos, computers, and other appliances that don't need to be plugged in.
- If you need to use an answering machine, turn off the ringer on your telephone. (Make sure you use a machine that allows you to receive and erase messages remotely.)
- If you don't need to use an answering machine, unplug all of your telephones.
- Ask the U.S. Postal Service to hold (or forward your mail), or have a trusted friend pick up your mail, newspapers, and other deliveries daily. Ask the friend to walk through your home periodically, making sure that your furnace, lights, timers, and so on, are working properly. (Your friend will have to be entrusted with your housekey.)

precautions. When a criminal recognizes you're a traveler, you become identified as someone who may be carrying money and valuables.

You can wear special clothing that allows you to hide your money and credit cards and is especially useful for thwarting pickpockets. With such clothing, you can keep most of your money hidden and carry only a few dollars in your purse or wallet. You want to be able to buy a newspaper or a snack without strangers seeing how much money you're carrying. Several companies make socks, belts, and lingerie with hidden compartments (see Figure 10–1).

Another useful item to carry is a whistle or similar noisemaker that can help you attract attention. The problem with a whistle is that its sound depends on your ability to blow into it. When you're scared, you may not be able to use it effectively. A personal alarm, such as the Quorum PAAL, can be easier to use and more dependable (see Figure 10–2). This small, electronic alarm can either be carried in the palm of your hand or attached to your belt buckle. When you sense a problem, you need only pull on its cord and a 104-db alarm is emitted. The alarm's noise should do two things: (1) surprise the attacker for a moment, giving you time to fight back or run, and (2) cause people nearby to look in your direction. The PAAL sells for about $30.

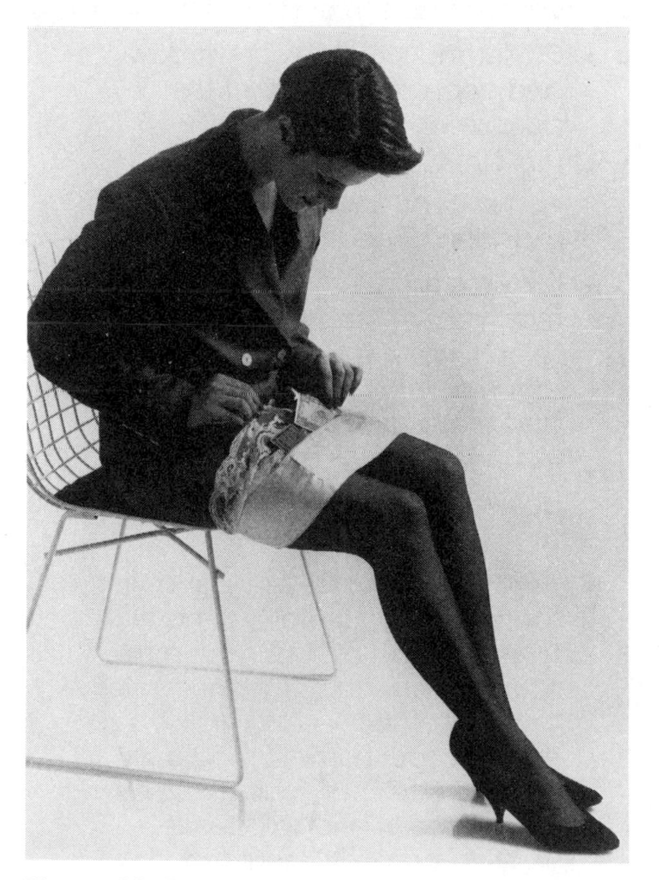

Figure 10–1. Clothing with hidden compartments can help you safeguard money and credit cards while traveling. (Courtesy of Hidden Assets)

Another useful electronic alarm can be attached to your property. The Quorum Elert is a self-contained unit with a built-in motion detector (see Figure 10–3). It can be attached to cameras, laptop computers, briefcases, luggage and other portable items, and it emits a 107-db alarm if someone tries to secretly move the items. The manufacturer's suggested list price for the Elert is about $50.

When you're in a hotel, you may want to place a personal alarm on your room door to alert you if someone tries to break in. Another useful device for room entrances is the brace lock (see Figure 10–4). It works on the same principle as a chair-back propped under a door knob to hold a door shut.

LEARNING FROM THE L.A. RIOTS

During the early evening of Wednesday, April 29, 1992, the eyes of the world were on Los Angeles as the worst riots in U.S. history erupted. The disturbance began within hours after a jury found four LAPD officers not guilty in connection with the videotaped beating of Rodney King.

The three days of rioting resulted in over 50 deaths and about three-quarters of a billion

Figure 10–2. A portable alarm allows you to attract attention at any time. (Courtesy of Quorum International, Inc.)

Figure 10–3. The "Elert" is a portable electronic alarm that can be attached to your luggage or other property. (Courtesy of Quorum International, Inc.)

dollars in damages. Using hindsight, we can see that there was much that the police, politicians, and community leaders could have done to prevent the riots. From those tragic days, much can be learned about staying safe in a riot. Unlike many of the victims who were caught by surprise and were forced to make split-second decisions, you can gain the advantage of being able to think calmly about how best to react.

On the first day of the disturbances, large mobs were on the streets, attacking motorists. Many drivers who saw the hostile groups immediately turned around and drove away. Some people who attempted to drive through the crowds were dragged out of their vehicles and beaten. Others had their windshields and windows smashed, or had their purses and packages snatched from the vehicles.

The obvious lesson is that it isn't a good idea to try to drive through a large, angry crowd. When in doubt about the intentions of a crowd, take the time to go another route to your destination. At best, you'll avoid entering a full-scale

Figure 10–4. A door brace can be useful when you're staying in a hotel. (Courtesy of MSI Mace)

Safety Tips for Your Home

- Keep working flashlights and a first-aid kit readily accessible.
- Keep a battery-operated radio (and fresh batteries) readily accessible.
- Program your telephones so that you can quickly dial the police and fire departments.
- Post emergency telephone numbers near each telephone.
- Post a diagram of fire escape routes and closest exits in each bedroom.

riot area; at worst, you'll miss a peaceful demonstration. As I explained in Chapter 9, the safest way to drive is with your doors locked and your windows rolled up.

If you're in your home and a riot is occurring outside, lock your doors and windows, close your shades and curtains, and call the police. If the disturbance is a large one, the police department's telephones will probably be busy. A busy line isn't bad news; it means that the police are probably aware of the situation.

In addition to trying to call the police, if you're in your home during a riot you should quickly gather your first-aid supplies, flashlights, a radio, batteries, and any weapons you may have. Be prepared for an electrical power outage.

If the disturbance is very close to your home, you should consider boarding up your windows from the inside (don't go outside). The boarding will prevent broken glass from flying around your home and will make it harder for a looter to break in.

RECOMMENDED RESOURCES

(See HM Sourcelist for more information.)

Hidden Assets—"Hidden-pocket" clothing

International Association of Home Safety and Security Professionals (IAHSSP)—Consumer information

Personal Security—Mace, security devices

Quorum International, Ltd.—Personal alarms

11

Insurance Considerations

By following the information given throughout this book, you'll greatly reduce the risk that your home will catch on fire or be broken into. However, unexpected disasters are always possible, and natural forces such as hurricanes, floods, lightning, or tornadoes could damage or destroy your home in seconds. Whether you rent or own your home, you need some kind of insurance. This chapter explains how to understand insurance policies, how to get the most coverage for your premium payments, and how to be sure your insurance company will pay you when you submit a legitimate claim.

Do you need homeowner's insurance? Ask yourself whether you would face financial hardship if you lost everything (or nearly everything) in your home, or if a visitor got hurt on your property and sued you for major medical expenses. For most people who have to work for a living, homeowner's insurance is a must.

UNDERSTANDING YOUR POLICY

Although some insurance companies print their policies on letterhead stationery, or introduce slight alterations of wording, most homeowners' policies are based on the HO series of policies developed by the Insurance Services Office. The policies that are most likely to be of interest to you are HO-1, HO-2, HO-3, HO-4, and HO-6. If you understand those policies, you should have no trouble identifying the versions that your insurance company uses.

All of the HO policies cover personal injury and property damage liability that you're found responsible for anywhere in the world; damage to your personal property; living expenses, if you can't live in your home because of a fire or other covered peril; and loss of credit cards.

The HO-1 is a basic policy that covers property against the following perils: fire or lightning; windstorm or hail; explosion; riot or civil disturbance; damage from aircraft or vehicles; smoke damage; vandalism or malicious mischief; theft; glass breakage; and volcanic eruption.

The HO-2 policy, called the "broad form," is similar to the HO-1, but adds coverage of the following perils: falling objects; freezing of plumbing, heating, or air-conditioning systems; weight of ice, snow, or sleet; sudden and accidental discharge from an artificially generated electric current; accidental discharge or overflow of water or steam from a plumbing, heating, or air-conditioning system; and sudden and accidental tearing apart, cracking, burning, or bulging of a heating, air-conditioning, or sprinkler system.

The HO-3 policy, called the "all-risk" policy, is the most popular. It covers the building (but not the contents) against every peril except those it specifically excludes. Three specific exclusions are: mine subsidence, earthquakes, and floods. The HO-3 covers the contents of a home only on a limited basis.

The HO-4 is a renter's insurance policy, and the HO-6 is used for condos and co-ops. Neither policy provides coverage for the building, but both cover personal property.

HOW MUCH BUILDING PROTECTION IS ENOUGH?

In addition to choosing the right coverage, you need to make sure you have designated the right dollar amount. Generally, a homeowner should have the building insured for at least 80 percent of its replacement cost—not its market value. It may not be necessary to have 100 percent replacement coverage, because there's little chance that your entire house will be lost. Even in a major fire, part of your house may be saved—and the insurance company won't pay you for that part. If you don't have coverage for at least 80 percent of the home's replacement value (building structure only: not contents), you may be severely penalized when you file a claim.

Your insurance company will pay a percentage of your damages based on the percentage of replacement cost stated in your policy. You have to maintain at least 80 percent of the replacement value of your home in order to receive 100 percent of the coverage amount stated in your policy. In other words, suppose your home's replacement value is $100,000 and you maintain $80,000 worth of coverage. If you were to suffer $40,000 worth of covered damages, the insurance company would pay you $40,000, or 100 percent of your loss. If you had maintained only $40,000 worth of coverage (half of 80 percent of the home's replacement value), the insurance company would pay you only $20,000, or 50 percent of the loss.

If you maintain coverage that represents 80 percent of your home's replacement value, you'll be self-insuring for the other 20 percent. Regardless of how much of a loss you suffer, the insurance company won't pay more than the amount of your policy.

Maintaining coverage for 80 percent of replacement value can be tricky because prices for materials change. You might want to have a little more coverage just to avoid a problem. It's a good idea to check the replacement value of your home periodically, to keep your coverage up-to-date. Some insurers offer "inflation guard" policies that automatically increase your coverage (as well as your premiums) based on the inflation rate.

PERSONAL OR HOME-BUSINESS PROPERTY

Your homeowner's policy will probably include a blanket amount of coverage for many personal items, but you may need additional coverage for some specific possessions. If you have an expensive piece of jewelry, for instance, your basic policy probably won't be enough. You might need to add a personal articles "floater" to the policy.

If you run a business from your home, you'll need to let your insurance agent know. Depending on the type of business you have, you might need special coverage for liability or business equipment and supplies. If you run a business from your home without informing your insurance agent, the insurance company may refuse to pay your claims.

SPECIAL DISASTERS

Depending on where you live, your home may be vulnerable to special disasters that may not be covered by a standard homeowners' policy. Residents of Arkansas, California, Illinois, Indiana, Kentucky, Mississippi, Missouri, New York, South Carolina, and the New England States are in danger of earthquakes.

In California, insurance premiums for earthquakes can cost up to $4 per $1,000 of coverage. In other places, the rates are as low as 30 cents per $1,000 of coverage. The problem with most earthquake insurance, however, is that the deductible may be as high as 15 percent.

Residents of Pennsylvania and other coal-mining areas are at special risk of mine subsidence damage when an underground mine collapses. Maps that show where coal was once mined are available in these regions. If you live near an abandoned mine, you should get coverage for mine subsidence damage. The State of Pennsylvania provides this type of insurance for a home, but not for the property inside.

Many people who face hurricanes and floods aren't fully covered for these two dangers. A standard homeowners' policy covers wind damage but not water damage. If a home is hit by a hurricane, the insurance company will pay claims only on the damage caused by the hurricane's wind, and not by the water that it carried to the house. Consider getting flood insurance if you live near a large body of water, or in an area with a high risk of hurricanes, or on land that has poor drainage and excessive runoff. You can obtain flood insurance through the National Flood Insurance Program. For information, contact the Federal Emergency Management Agency (FEMA).

AVAILABLE DISCOUNTS

When shopping for homeowners' insurance, be aware that you may be able to take advantage of one or more premium discounts. Different insurance companies give different discounts, so you may have to shop around to get the best price. Here are some things that discounts are given for:

A newly built home;

Being age 55 or over;

Being a nonsmoker;

Having no prior losses;

A burglar alarm;

Deadbolt locks on all exterior doors;

Smoke detectors;

A sprinkler system.

In addition to using all applicable discounts, you may be able to save money by getting your homeowners' policy from the same company that insures your car. Another way to save money is by paying your premiums annually rather than quarterly (with a service charge added). You may also want to consider lowering your premiums by raising your deductible. However, don't raise it so high that a loss will be too much of a financial burden.

Reconsider Your Insurance Needs If:

- You install a burglar alarm;
- You upgrade your home's wiring or plumbing:
- You install a sprinkler system;
- You buy a computer, jewelry, furs, or expensive electronic equipment (you may need a rider);
- You start a business at home;
- You make any substantial home improvements that increase your home's replacement value;
- You retire (you may become eligible for a discount).

FEDERAL CRIME INSURANCE PROGRAM

Because of increasing crime rates, homeowners and tenants have sometimes been unable to find affordable crime insurance through private insurers. In response to the problem, legislation was passed in 1970 to create federal crime insurance. Today, the Federal Insurance Administration, as part of the Federal Emergency Management Agency (FEMA), offers low-cost crime insurance to residents and business owners in certain states.

You can obtain federal crime insurance if you live in: Alabama, California, Connecticut, Delaware, Florida, Georgia, Illinois, Kansas, Maryland, New Jersey, New York, Pennsylvania, Rhode Island, Tennessee, the District of Columbia, Puerto Rico, or the Virgin Islands.

Federal crime insurance covers losses from burglary and robbery, including damage done to a home or business during the crime. It does not cover mysterious disappearances or thefts of property that were not observed by the insured.

Residents can purchase the insurance in increments of $1,000, up to a maximum of $10,000. The annual premiums vary, depending on the amount of coverage and whether the insured is protected by a burglar alarm system. Premiums for 1992 were:

Amount of Coverage	Annual Premium without Alarm Credit	Annual Premium with Alarm Credit
$ 1,000	$ 32	$ 30
2,000	42	40
3,000	52	50
4,000	62	58
5,000	74	70
6,000	84	80
7,000	94	90
8,000	104	98
9,000	116	110
10,000	126	120

Losses, subject to the deductible and applicable limitations, are payable up to the full

O.M.B. NO. 3067-0031
Expires March 31, 1994

FEDERAL EMERGENCY MANAGEMENT AGENCY
FEDERAL INSURANCE ADMINISTRATION

CRIME INSURANCE SWORN STATEMENT AND PROOF OF LOSS

DATE OF LOSS	POLICY NUMBER	CLAIM NUMBER

1. Claim is hereby made for $ ____________ loss, as fully described in the itemized statement of loss following, under Policy No. ____________ Issued to ____________
   With premises at ____________
2. On the ______ day of ____________, 19__, about ______ m, a loss occurred at ____________ *(Number)*
   ____________ *(Street)* *(City)* *(County)* *(State)*
   Occurrence of the loss was first known to the Insured on the ______ day of ____________, 19__, with oral notice given to the Insurance Agent, Broker or Servicing Company on the ______ day of ____________, 19__.
3. The nearest Law Enforcement Authorities were notified on the ______ day of ____________, 19__, and a report was submitted to the following Officers: ____________ *(Copy of report to Law Enforcement Authority should be attached, if available.)*
4. The loss was sustained as follows: *(Briefly state the facts of how the loss occurred. In the case of robbery, state where and how the crime occurred. In the case of burglary, state location and nature of entry to premises, and to any insured safe or vault. If damage to insured property occurred, describe how such damage was sustained. Name persons known or suspected as implicated.)*
   ____
5. The property insured on which loss is claimed belonged at the time of loss to ____________
   and no other person had any interest therein except as follows: ____________
6. No other insurance was carried on this property by the Insured at the time of loss except as follows: ____________
7. The Insured never before suffered loss from a similar cause, nor received indemnity therefor except as follows: ____________
8. No material fact regarding the loss has been suppressed, withheld, or misrepresented herein. Any information that may be required by the Insurer will be furnished on demand and will be considered a part of the Sworn Statement and Proof of Loss.
9. It is understood and agreed that submission of this Sworn Statement and Proof of Loss form to the Insured or its preparation by a representative of the Insurer or the acceptance or retention of the Sworn Statement and Proof of Loss by the Insurer is not a waiver of any rights under the Policy.
10. The following is a complete statement of the extent of said loss: *(Describe each item in detail. Include serial or identification numbers or markings. If precious stone, state weight and quality of stone, style and quality of mounting. Attach relevant receipts, bills, and other supporting papers. If schedule following is not suitable, describe additional items on a separate paper and attach.)*

FEMA Form 81-46, OCT 91   REPLACES ALL PREVIOUS EDITIONS

Figure 11–1. Before you need to file a claim, find out what the federal crime insurance form will expect of you. (Courtesy of Federal Emergency Management Agency, Federal Insurance Administration)

DESCRIPTION OF PROPERTY OR DAMAGE	NAME OF OWNER	WHEN AND WHERE PURCHASED OR OBTAINED	ACTUAL COST	DEPRECIATION IN VALUE	AMOUNT CLAIMED

The said loss did not originate by any act, design or procurement on the part of this insured: nothing has been done by or with the privity or consent of this insured to violate the conditions of the policy or render it void; no articles are mentioned herein or in annexed schedules but such as were taken or damaged at the time of said loss; no property has in any manner been concealed, and no attempt to deceive the Insurer, as to the extent of said loss, has in any manner been made. Any other information that may be required will be furnished or considered as part of this proof.

The insured hereby covenants that no release has been or will be given to or settlement or compromise made with any third party who may be liable in loss or damages to the insured and the insured in consideration of the payment made under this policy hereby subrogates the said Insurer to all rights and causes of action the said insured has against any person, persons or corporations whomsoever for loss or damage arising out of or incident to said loss or damage to said property and authorizes said Insurer to sue in the name of the insured but at the cost of the Insurer any such third party, pledging full cooperation in such action.

The furnishing of this blank or the preparation of proofs by a representative of the above Insurer is not a waiver of any of its rights.

CERTIFICATION BY INSURED: *"I certify, under penalty of Federal law for fraud or misrepresentation as set forth in 18 U.S.C. 1001, that the statements I have made in this Proof of Loss are true and correct to the best of my knowledge and belief."*

Insured ______________________________ Date ______________

State of ______________________ County of ______________________

Suscribed and sworn to before me this ________ day of ______________, 19 ______

______________________________ Notary Public.

Figure 11–1. *Continued.*

amount of the policy coverage for each separate occurrence. However, the limit of liability for loss or damage in any one occurrence can't exceed the cost of repairing or replacing the property with other property of like kind and quality; actual cash value at the time of loss; or the limit of insurance stated in the application.

Each loss is subject to a minimum deduction of $100. There is an aggregate limit of $1,500 per occurrence or $500 for any one article of jewelry (including watches, necklaces, bracelets, rings, gems, and precious and semiprecious stones); any articles of gold, silver, or platinum (including flatware and holloware); furs, fine arts pieces, antiques, and coin and stamp collections. There is also a $200 cash limit and a ceiling of $500 for lost securities.

Tenants who aren't related to the insured or to a permanent member of the insured household must purchase a separate policy to be insured. (The $10,000 limit of coverage per house or apartment can't be increased by subdividing the premises among the occupants.)

To be eligible for federal crime insurance, you must meet certain protective-device guidelines. All of your exterior doors except sliding glass doors must have a deadbolt, jimmyproof deadlock, or self-locking deadlatch lock. Except as may be prohibited by local fire ordinances, all exterior sliding glass doors and windows that open into stairways, porches, platforms, or other areas that allow access to the home must be equipped with some type of locking device.

With federal crime insurance (or with other homeowners' insurance), always be prepared to file a claim. Learn exactly what you'll need to show in order to receive coverage (see Figure 11–1).

RECOMMENDED RESOURCES

(See HM Sourcelist for more information.)

Federal Crime Insurance Program—Federal agency

12

Choosing and Using Security Professionals

Although there is much you can do to protect yourself, sometimes it's best to hire security professionals. You would probably save time and money by getting a locksmith to re-key the locks in your home, for example, rather than trying to re-key them yourself. When looking for security services, most people randomly pick a company from a local telephone book (see Figure 12–1). But beware: too many security-related businesses are run by dishonest, self-proclaimed "experts" who have had little security training or experience. At best, relying on an incompetent or dishonest security professional can cost you money unnecessarily. At worst, it can place you and your family in danger. This chapter explains what various security professionals can do for you, how to find good ones, and how to get the most out of them.

Using Your Telephone Directory to Find Security Products and Services

PRODUCT/SERVICE	"YELLOW PAGES" HEADINGS
Burglar Alarms/Installers	Burglar Alarm Systems Burglar Alarms & Detection Devices
Burglar Bars	Building Materials – Retail Burglar Bars Doors Hardware – Builders Hardware – Retail Locks & Locksmiths
Car Alarms/Installers	Automobile Alarm Systems Automobile Radios & Stereo Systems Sales and Service Burglar Alarm Systems Burglar Alarms & Detection Devices Locks & Locksmiths Stereophonic & High Fidelity Equipment Dealers
CCTVs/Installers	Burglar Alarm Systems Burglar Alarms & Detection Devices Security Control Equipment & Systems
Doors	Building Materials – Retail Burglar Bars Doors Hardware – Builders Lumber – Retail Storm Windows & Doors Windows
Door/Window Hardware	Building Materials – Retail Burglar Bars Doors Hardware – Retail Hardware – Builders
Fire Extinguishers	Fire Extinguishers Hardware – Retail Smoke Detectors & Alarms
Home Alarms/Installers	Burglar Alarm Systems Burglar Alarms & Detection Devices Locks & Locksmiths Security Control Equipment & Systems
Home Automation/Installers	Burglar Alarm Systems Burglar Alarms & Detection Devices Controls, Control Systems & Regulators
Locks/Locksmiths	Building Materials – Retail Hardware – Retail Locks & Locksmiths
Safes/Installers	Hardware – Retail Locks & Locksmiths Safes & Vaults
Safety/Security Consultants	Burglar Alarm Systems Burglar Alarms & Detection Devices Crime Prevention Programs Locks & Locksmiths Security Consultants Security Control Equipment & Systems
Security Glass	Glass – Automobile, Plate, Window, etc. Glass – Plate, Window, etc. Windows
Security Lighting	Building Materials – Retail Hardware – Retail Lamps & Lamp Shades – Retail Lighting Fixtures – Retail
Self-Defense Products/Training	Judo, Karate & Kung-Fu Schools Judo, Karate & Kung-Fu Supplies & Equipment Karate & Other Martial Arts Instruction Karate & Other Martial Arts Supplies & Equipment Police Equipment Protective Devices Self-Defense
Smoke Detectors	Building Materials – Retail Hardware – Retail Smoke Detectors & Alarms
Windows	Building Materials – Retail Doors Lumber – Retail Storm Windows & Doors Windows

IAHS &SPI™
INTERNATIONAL ASSOCIATION OF HOME SAFETY AND SECURITY PROFESSIONALS

Figure 12–1. Specialists in home protection will be listed under many headings in your local Yellow Pages. (Used with permission of the International Association of Home Safety and Security Professionals, Inc. (IAHSSP))

In addition to locksmiths, the security professionals you might use include security consultants, self-defense instructors, and installers of home alarms, home automation systems, and car alarms. Before hiring any service provider, make sure that the company or individual has been doing business in your area for a few years and doesn't have a trail of unhappy customers. You can often get that information by calling your Better Business Bureau, state Attorney General's office, Consumer Protection Agency, or local Chamber of Commerce. If the company is a corporation, you can find out when it began doing business by calling your state Attorney General's Corporation Department.

Before hiring a security professional, you'll also need to know if the person is competent. The best way to do that is to consider his licenses, certifications, memberships, training, experience and status among his professional peers. The questions you'll need to ask vary among types of security professionals.

LOCKSMITHS

Locksmiths primarily sell and service locking devices, safes, and other security hardware. Some also handle home and car alarms. If you just need someone to unlock your car door, you probably don't need to worry about which locksmith you choose. Virtually all locksmiths know how to open most cars. To avoid being charged too much when you are locked out of your car, don't call only the first locksmith listed in your telephone book. Call two or three and ask each one what the charge will be. Stay away from locksmiths who won't give you a flat fee. The charge should depend on how far away your car is from the locksmith's shop, and on your car's make, model, and year—not on how long it takes to unlock your door. Locksmiths who have different skill levels will take different amounts of time to unlock the same vehicle, but you shouldn't have to pay more to a locksmith who isn't very skillful! (Expect to pay more, however, if new car keys must be made for you.)

Be sure to check out carefully any locksmith who will work at your home. A small percentage of them are former burglars. In most places, a locksmith doesn't need to be licensed. Even where "licensing" is required, an applicant may not have to submit to a criminal background check or meet any competency or training standards. The applicants usually need only sign a business registration form and pay a small fee. New York City and California are among the few places that require applicants for a locksmith license to submit to a background check and meet certain training and competency standards.

Where licensing laws are lax or nonexistent, you'll need to consider a locksmith's certifications. The most respected locksmith certification program is offered by the Associated Locksmiths of America (ALOA), the oldest and largest nonprofit locksmith trade association. ALOA offers three certification levels to members and nonmembers: Registered Locksmith (RL), Certified Professional Locksmith (CPL), and Certified Master Locksmith (CML). If a locksmith's Yellow Pages advertisement includes ALOA's logo and the initials RL, CPL, or CML, the locksmith is showing certification by the ALOA. Some locksmiths simply advertise themselves as "Master Locksmiths." Don't pay attention to the term because it has no meaning.

Don't be surprised if you can't find an ALOA-certified locksmith near you. Most locksmiths—including many highly respected ones—haven't been tested by ALOA. Some disagree with the organization's legislative activities; others think the testing fees are too high, or don't believe testing is necessary. It's important to realize that, although ALOA certification is an indication of competency, lack of such certification doesn't necessarily indicate incompetency.

Manufacturers of high-security locks and cylinders also offer certification programs for locksmiths. Although these certifications are based on proficiency with a specific product line, they also indicate overall competency because a locksmith has to have a good foundation of knowledge to understand high-security locks.

Next to certification, the best indicator of competency for a locksmith is membership in certain associations that provide continuing education to their members. You may have seen the various logos in your Yellow Pages under "Locks & Locksmiths." The best-known national associations for locksmiths are: ALOA, Door and Hardware Institute, and The National Locksmith Association and Trust. ALOA and the Door and Hardware Institute (DHI) have tough membership criteria; the National Locksmith Association (NLA) and Trust are basically just magazine subscriptions. Their only requirement for "membership" is a subscription to a trade journal. Nevertheless, the publications help keep locksmiths abreast of fast changing technical and product information.

In addition to national associations (which include members from both the United States and Canada), there are many state and regional locksmith associations—British Columbia Locksmiths Association, Central Pennsylvania Locksmith Association, Locksmith Association of Connecticut, Maryland Locksmiths Association, Master Locksmiths of Quebec, Missouri-Kansas Locksmith Association, Oregon Association of Professional Locksmiths, and many others. The groups vary greatly in their membership criteria and educational benefits. Still, a locksmith who is a member of any of the major locksmith associations is a better risk than someone who doesn't belong to any of them. Generally, if a locksmith belongs to one of the groups, the membership will be stated in the telephone directory display advertisement.

A locksmith who specializes in safes may also belong to the Safe and Vault Technicians' Association (SAVTA) or the National Safeman's Organization. SAVTA is the more prestigious of the two and is harder to join.

Don't be surprised if you find that few locksmiths in your area are members of any association. The majority of locksmiths have never joined an association. As with certifications, lack of membership doesn't indicate incompetency, but lack of proper training and experience does.

Locksmiths generally learn their trade in apprenticeship training, correspondence courses, or classroom training. A well-staffed, accredited school provides the broadest base of knowledge. Apprenticeship training can be great, if done under a competent locksmith. Correspondence courses are hard to judge: a person can pass the tests (mainly written exams) without learning much. A locksmith who graduated from an accredited school is a better risk than one who didn't.

Next in importance to how locksmiths learned the trade is how long they've been practicing it. Look for a locksmith who has been working in your area for at least 5 years.

HOME-ALARM SYSTEMS INSTALLERS

Alarm installers generally handle a wide range of electronic security devices, including home alarms, home automation systems, and access control systems. Some also install car alarms.

As with locksmiths, alarm installers should be carefully checked out before they work in your home. The person who installs your electronic security system can easily disable it at any time (or tell someone else how to disable it). Also, a lot can go wrong with an improperly installed system. You'll save yourself and your neighbors a lot of sleepless nights by choosing a competent person to install your system.

Many cities and states license alarm installation companies—that is, the company itself or the owner is often required to be licensed, but the person actually doing the work may not be licensed. Call your state and local licensing agencies to learn about the requirements in your area.

Regardless of the local licensing requirements, both the company and the installer should have been installing security systems for at least 5 years. Neither should have a bad record with your local or regional Better Business Bureau.

You can be reasonably sure that you are hiring a competent person if your alarm installer has been certified by the National Alarm Association of America (NAAA) or the National Burglar and Fire Alarm Association (NBFAA). Both organizations offer highly respected certification programs for members and nonmembers. In addition to those two national associations, there are many state and regional groups for alarm installers. As with locksmiths, however, lack of certifications or memberships doesn't necessarily indicate incompetency.

Alarm systems installers usually learn their trade through apprenticeship or classroom training. In most places that license alarm installers, you don't need to worry about where your installer learned the trade. If you want the best alarm installer around, look for one who has been certified by a national trade association *and* is a licensed electrician.

PHYSICAL SELF-DEFENSE INSTRUCTORS

Any physical self-defense instructor should have a black belt (or equivalent training) in one or more martial arts, and at least 3 years' experience as an instructor. Certifications and memberships are less important considerations when choosing a self-defense instructor than when choosing other security professionals.

Don't be impressed by how many trophies and plaques an instructor has. Those awards simply mean that he or she has won some martial arts competitions. You're looking for streetwise self-defense, and a martial arts competition no more resembles a street encounter than does a boxing or wrestling match. On the street, you have no padded protective gear, no referees, no timed rounds, and no rules. You want an instructor who knows how to fight and how to teach fighting—not necessarily one who knows how to compete.

Look for someone who emphasizes self-defense techniques that a smaller person can use to defend against a larger person. These include kicks below the waist, open-handed strikes to vulnerable areas of the body, and the use of weapons. Be suspicious of instructors who teach "Bruce Lee style" high kicks. They look great in the movies, but high kicks seldom work well in a street fight. Low kicks provide better balance and more speed and power. (If the instructor is teaching high kicks mainly as an exercise in balance and flexibility, that's fine. But if you are told that such kicks are good for self-defense, find another instructor.) Stay away from instructors who emphasize hardening of the hands and feet, breaking boards with your limbs, or brute-strength fighting techniques. These feats are impractical for most people and take many years to learn.

The best way to judge a physical self-defense instructor is to observe one during a class. All good instructors will be happy to let you watch at no charge. Pay attention to how the instructor interacts with the students. Ask yourself: Do they respect the instructor? Do they understand the instructions? Is the class run in an orderly fashion? Does the instructor know how to fight?

SECURITY CONSULTANTS

Although this book contains everything most people need to know about home security, there's much more to the subject than can be included in one book. If you live in an expensive house, or keep high-priced valuables in your home, or have a high-profile job, or know that people want to harm you or your family, you should consult with a qualified security consultant. This professional can help you to assess your security risks accurately and can show you the most cost-effective ways to meet those risks.

Many locksmiths and home-alarm installers who call themselves "security consultants" are basically salespersons. Because they mainly want to sell you something, they're likely to suggest products and services that you don't need. In some cities, police officers give free security advice. However, most police officers aren't specialists in using security hardware or electronic alarm systems. The advice they give is very general: "Always use deadbolts"; "Keep your doors locked"; "Keep your windows closed."

For unbiased, accurate, and in-depth advice, you'll need an independent security consultant, someone who has a broad range of security-related training and experience and can show you ways to get the best protection at the lowest cost. Independent security consultants know where to find a wide variety of security products at discount prices. Often, they can save you more money than they charge. But again, beware: not all independent security consultants are useful.

In most places, anyone can legally call himself or herself a security consultant. No special training or licensing is required. A competent security consultant will have had extensive training and experience in two or more of the following areas: criminology, law enforcement, locksmithing, alarm systems installation, safe servicing, martial arts, or fire fighting. To find out what training or experience a security consultant has, *ask*. He or she should be happy to provide the information to you in writing (usually in the form of brochures and flyers).

Try to get references before you hire a security consultant. That may be hard because of the confidential nature of the profession. It's not in the best interest of most clients to act as reference sources.

Another indication that a security consultant is qualified is whether he or she has been certified by a security trade association. The most prominent associations are the American Society for Industrial Security (ASIS), the International Association of Home Safety and Security Professionals (IAHSSP), and the International Association of Professional Security Consultants (IAPSC).

ASIS is the oldest and largest trade association for security consultants. It offers the Certified Protection Professional (CPP) certification to members and nonmembers who meet certain guidelines and pass a rigorous test. IAHSSP is the only association that specializes in training security professionals who serve homeowners and apartment dwellers. It offers the Residential Protection Specialist (RPS) certification to members who pass a rigorous test. The IAPSC offers the Certified Protection Officer (CPO) status to members and nonmembers who pass a test.

A friend asked me: "If I didn't know you, how would I find the best security consultant around?" That question made me seriously consider how a layperson could distinguish among a group of competent security consultants. A comparison of licenses, certifications, memberships, training, and experience probably wouldn't help much, because several of the consultants might appear to be equally qualified. To find top-notch security consultants, you have to consider their professional reputations among their peers. The most respected consultants are those who hold or have held offices in trade associations, or teach security classes at accredited schools and trade association seminars, or write books for the trade, or have technical articles published in the major trade journals. Unless you live in a major city, it's unlikely that more than one security consultant in your area will meet one of those criteria. The person who has those credentials is probably the best around.

RECOMMENDED RESOURCES

(See HM Sourcelist for more information.)

International Association of Home Safety and Security Professionals (IAHSSP)—Trade Association

13

The Safe and Secure Home

Throughout this book, we've covered lots of information about safety and security systems, devices, and hardware. We've discussed the importance of getting homeowners' insurance and of using psychological self-defense. If you've read all the other chapters, you know more about home security than most burglars do, and you have all the information necessary to think like a security professional.

This chapter discusses how you can put everything together to make your home as safe as you want it to be. It's important to understand that no single security plan is best for everyone. Each home and neighborhood has unique strengths and vulnerabilities, and each household has different needs and limitations.

The important limitation most of us face is money. If money were no object, it would be easy for me to lay out a great security plan for you. I would advise you to build a steel frame construction Smart House with steel doors, multiple-bolt locks, high-security cylinders, bullet-resistant glass, and an integrated CCTV system. I might also advise you to hire armed guards. Those purchases would make your home more secure, but they would be overkill.

With proper planning, you can be safe in your home without spending more money than you can afford or being too inconvenienced. Proper planning is based on the following considerations:

How much money are you willing to spend?

How much risk is acceptable to you?

How much inconvenience is acceptable to you?

How much time are you willing to spend on making your home more secure?

How much of the work are you willing to do yourself?

No one can answer those questions for you. You are the best person to design a security plan that meets your needs.

Before you can create the best plan for your home, you'll need to conduct a safety and security survey (or "vulnerability analysis). The survey requires your walking around the outside of your home and through every room. You should make note of all potential problems.

SURVEYING A HOUSE

The purposes of a safety and security survey are:

To help you identify potential problems;

To assess how likely and how critical each risk is;

To determine cost-effective ways to either eliminate the risks or bring them to an acceptable level.

The survey will allow you to take precise and integrated security and safety measures.

A thorough survey involves not only inspecting the inside and outside of your home, but also examining all of your safety and security equipment and reviewing the safety and security procedures used by all members of your household. The actions people take (or fail to take) are just as important as the equipment you may buy. What good are high-security deadbolts, for instance, if residents often leave the doors unlocked?

As you conduct your survey, keep the information in the preceding chapters in mind. You'll notice many potential safety and security risks (every home has some). Some of the risks will be simple to reduce or eliminate immediately. For others, you'll need to compare the risk to the cost of properly dealing with it. There's no mathematical formula to fall back on. You'll need to make subjective decisions based on what you know about your household and neighborhood and guided by the information throughout this book.

When surveying your house, it's best to start outside. Walk around the house and stand at the vantage points that passersby are likely to have. Many burglars will target a home because it's especially noticeable while driving or walking past it. When you look at your home from the street, note any feature that might make someone think that no one is home, or that a lot of valuables may be in the home, or that the home may be easy to break into.

Keep in mind that burglars prefer to work in secrecy. They like heavy shrubbery or large trees that block or crowd an entrance, and they like homes that aren't well-lighted at night. Other things that may attract burglars' attention include expensive items that can be seen through windows, a ladder near the home, and notes tacked onto the doors.

As you walk around the home, note anything that might help discourage burglars. Can your "Beware of Dog" sign or your fake security system sticker be seen in the window? Walk to each entrance and consider what burglars might like and dislike about them. Is the entrance well-lighted? Can neighbors see someone who's at the entrance? Is there a video camera pointing at the entrance? Does the window or door appear to be easy to break into?

After surveying the outside of your home, go inside and carefully examine each exterior door, window, and other opening. Consider whether each one is secure but allows you to get out quickly. Check for the presence of fire safety devices. Do you have enough smoke detectors and fire extinguishers? Are they in working order? Are they in the best locations?

Take an honest look at the safety and security measures that you and your family take. What habits or practices might you want to change?

HOME SAFETY AND SECURITY CHECKLIST

Because every family, home, and neighborhood is unique, no safety and security survey checklist can be comprehensive enough to cover all of every home's important factors, but the following checklist will help to guide you during your survey. Keep a notepad handy to write down details or remedies for potential problems.

When you've finished reading this book, keep it handy and place a bookmark at pages 162–63, where the Home Safety and Security Checklist appears. If you are intent on managing your home security to your best advantage, you'll review the checklist periodically, to be sure that you've responded to your changing security needs.

SURVEYING AN APARTMENT

In many ways, surveying an apartment is like surveying a house. The difference is that you have to be concerned not only about the actions of your household, but also about those of your landlord, the apartment managers, and the other tenants. The less security-conscious others around you are, the more at-risk you will be. No matter how much you do to avoid causing a fire, for example, a careless neighbor may cause one. If your neighbors don't care about crime prevention, your apartment building or complex will be more attractive to burglars.

As you walk around the outside of your apartment, notice everything that would-be burglars might notice. Will they see tenants' "crime watch" signs? Will they see that all the apartments have door viewers and deadbolt locks? Burglars hate a lot of door viewers because they never know when someone might be watching them.

After surveying the outside, walk through your apartment and look at each door, window, and other opening. If you notice major safety or security problems, point them out to your landlord. You might also want to suggest little things that the landlord can do to make your apartment more secure.

High-Rise Apartment Security

High-rise apartments have special security concerns that don't apply to apartments with fewer floors. In a high-rise, more people have keys to the building, which means more people can carelessly allow unauthorized persons to enter.

The physical structure of a high-rise often provides many places for criminals to lie in wait for victims or to break into apartments unnoticed. Many high-rise buildings aren't designed to allow people to escape quickly during a fire.

The safest apartments have only one entrance for tenants to use, and that entrance is guarded 24 hours a day by a doorman. An apartment that doesn't have a doorman should have a video intercom system outside the building. Video intercoms are better than audio intercoms because they let you see and hear who's at the door before buzzing the person in.

Home Safety and Security Checklist

As you conduct your survey, note each potential problem that is of concern to you.

Home Exterior

- ☐ Shrubbery. (Shouldn't be high enough for a burglar to hide behind—or too near windows or doors.)
- ☐ Trees. (Shouldn't be positioned so a burglar can use them to climb into a window.)
- ☐ House numbers. (Should be clearly visible from the street.)
- ☐ Entrance visibility. (Should allow all entrances to be seen clearly from the street or other public area.)
- ☐ Lighting near garage and other parking areas.
- ☐ Ladders. (Shouldn't be in the yard or in clear view.)
- ☐ "Alarm System" or "Surveillance System" stickers. (Shouldn't identify the type of system that's installed.)
- ☐ Mailbox. (Should be locked or otherwise adequately secured, and should show no name or only a first initial and last name.)
- ☐ Windows. (Should be secured against being forced open, but should allow for easy emergency escape.)
- ☐ Window air conditioners. (Should be bolted down or otherwise protected from removal.)
- ☐ Fire escapes. (Should allow for easy emergency escape but not allow unauthorized entry.)

Exterior Doors and Locks

(Included here are doors connecting a garage to the home.)

- ☐ Door material. (Should be solid hardwood, fiberglass, PVC plastic, or metal.)
- ☐ Door frames. (Should allow doors to fit snugly.)
- ☐ Door glazing. (Shouldn't allow someone to gain entry by breaking it and reaching in.)
- ☐ Door viewer (without glazing). (Should have a wide-angle door viewer or other device to see visitors.)
- ☐ Hinges. (Should be either on inside of door or protected from outside removal.)
- ☐ Stop molding. (Should be one-piece or protected from removal.)

- [ ] Deadbolts. (Should be single-cylinder with free-spinning cylinder guard and a bolt with a 1-inch throw and a hardened insert.)
- [ ] Strike plates. (Should be securely fastened.)
- [ ] Door openings (mail slots, pet entrances, and other access areas). (Shouldn't allow a person to gain entry by reaching through them.)
- [ ] Sliding glass doors. (Should have a movable panel mounted on interior side, and a bar or other obstruction in the track.)

Inside the Home

- [ ] Fire extinguishers. (Should be in working order and mounted in easily accessible locations.)
- [ ] Smoke detectors. (Should be in working order and installed on every level of the home.)
- [ ] Rope ladders. (Should be easily accessible to bedrooms located above the ground floor.)
- [ ] Flashlights. (Should be in working order and readily accessible.)
- [ ] First-aid kit. (Should contain fresh bandages, wound-dressing and burn ointments, aspirin, and plastic gloves.)
- [ ] Telephones. (Should be programmed to dial the police and fire departments quickly or their phone numbers should be posted nearby.)
- [ ] Burglar alarm. (Should be in good working order and adequately protected from vandalism, and should have adequate backup power.)
- [ ] Safes. (Should be installed so they can't be seen by visitors.)

People in the Home

- [ ] Doors (locking). (Should be locked by all residents every time they leave the home—even if they plan to be gone for only a few minutes.)
- [ ] Doors (opening). (Should not be opened by any resident unless the person seeking entry has been satisfactorily identified.)
- [ ] Fire escape plan. (Should be familiar to all residents as a result of practicing how to react during a fire.)
- [ ] Confidentiality. (Should be maintained by all residents regarding locations of safes, burglar alarms, and other security devices.)
- [ ] Money and valuables. (Should be stored in the home only in small amounts and only if they have low resale or "fencing" value.)
- [ ] Drapes and curtains. (Should be routinely closed each night.)
- [ ] Garage doors. (Should be kept closed and locked.)

It's also important to consider the building's fire exits. Every fire exit door leading directly to apartments should stay locked from the fire exit side. In other words, a person who's in a fire exit stairwell should have to walk out of the building and re-enter through the main entrance to get onto any floor. The fire exits should be only for getting out—not for getting in.

All the fire exit doors on the main floor should be connected to an alarm that will sound if someone opens one of the doors and that can be heard by the doorman (or by a maintenance person).

Exterior fire escapes—the metal stairs on the outside of a building—present another problem. Many of them have ladders that can be reached from the ground. Burglars can use a stick to disengage the hook that holds the ladder, and make the ladder slide down. They can then climb the ladder to get into any apartment accessible via the fire escape. Apartments that can be accessed from an exterior fire escape should have grates inside that will prevent anyone from climbing through the windows. All exterior doors leading directly to a fire escape should be strong and equipped with good locks.

Elevators should be programmed so that they always stop at the ground floor before descending to the basement and before going to the upper floors. Too often, an elevator passenger is overcome by an attacker and taken directly to the basement or to another floor to be assaulted. If the elevator always stops at the ground floor, a captive has more of a chance of getting out or of being seen.

Elevators should have corner-mounted mirrors that allow you to see who's on the elevator before you get on. The mirrors should be positioned so that there is no hiding spot in the elevators. If you see someone who makes you feel uncomfortable, you can avoid getting onto the elevator.

As with any type of apartment, you should consider how security-conscious your neighbors are. In a high-rise, security will depend largely on whether the building's management stresses the importance of security to the tenants and reprimands flagrant violations.

One problem in many high-rises is that burglars are let into the building by a tenant who doesn't bother checking the identity of people before buzzing them in. Some burglars get into buildings by "piggybacking"—running in behind a tenant who has a key to the door. Many tenants don't like to question a stranger who piggybacks for fear of being attacked.

FINAL THOUGHTS

Wouldn't it be nice if you could create a security plan and then never again need to think about security? I'm sure you realize that continued security requires your continued commitment—not only always locking doors behind you, but also staying abreast of security developments from which you may be able to benefit.

To learn about the latest security products, and for a continuous supply of money-saving ways to stay safe, I suggest you read *Home Mechanix*. It's the only home improvement magazine with a "Home Security" column. Contributors to the column include various security specialists, the editors of *Home Mechanix*, and yours truly.

If you have any questions or comments about this book, let me hear from you. You can write to me at this address:

Bill Phillips
c/o IAHSSP
P.O. Box 2044
Erie, Pennsylvania 16512-2044

Appendix A

Ways to Use X-10 Products*

Most home alarm and home automation systems allow you to use X-10 modules to control your appliances and lights automatically. Because the modules are sold separately, you can use them to upgrade or custom-design your system. The following pages show a few of the many ways to use them.

---

*Courtesy of X-10 (USA) Inc.

Introduction

The *X-10* ® *POWERHOUSE* ™ Home Control System can deter intruders by making your home look and sound lived in whether or not you are there. You never have to come home to a dark house again, or leave the outside lights on all day to get the key in the door at night. The Home Automation Interface, Mini Timer or Clock Radio/Timer can turn the outside lights on for you... Automatically! The Wireless Transmitter and Transceiver set lets you turn lights on from the safety of your car as you pull into your driveway. The *SUNDOWNER* ™ will even turn lights on automatically when the sun goes down and turn them off again at dawn... with no programming!

The *X-10 POWERHOUSE* system can wake you up to stereo or TV news, light up your bedroom, hallway, bathroom; start the coffee, start your central heating or air conditioning, warm up the curling iron, and does it all before you're even out of bed. At night the system can lower the heat, play music or your favorite late night TV show for as long as you set it, and can later turn off the lights automatically.

There are many types of Controllers available including the Mini Timer, Mini Controller, Maxi Controller, Clock Radio/Timer, Telephone Responder, Home Automation Interface (for computers), Burglar Alarm Interface etc. These Controllers simply plug in and send signals over your existing house wiring to plug-in Modules.

Many types of Modules are available including: Appliance Modules for TVs, stereos, coffee pots etc. Lamp Modules which contain dimmers and can be used for incandescent lamps up to 300 watts. Wall Switch Modules which also contain dimmers and can be used for incandescent outdoor lights and ceiling lights up to 500 watts. Heavy Duty Appliance Modules control 220V air conditioners and water heaters. The Thermostat Set-back Controller handles central heating and air conditioning. The 3-Way Wall Switch Module controls incandescent lights operated from two switches. The Split Receptacle Module replaces an existing wall outlet.

For safety

Keep in mind that *X-10 POWERHOUSE* Controllers will always turn lamps and appliances on or off the instant you press the buttons or at the times you have programmed them to do so. That's obvious - but there can be some unexpected consequences.

For example, an empty coffee pot can be remotely turned on. If that should happen, your coffee pot may be damaged from overheating. If an electric heater is turned on by remote control while clothing just happens to be draped over it, a fire could result.

Therefore, always be aware of what appliance you are turning on or off so that potentially dangerous situations will not occur.

How it works

The *X-10 POWERHOUSE* System controls lights and appliances throughout your home from any convenient location. The Controllers transmit signals to the Modules over your existing house wiring. You plug lamps into Lamp Modules, plug appliances into Appliance Modules and replace wall light switches with Wall Switch Modules. You can then control virtually everything electrical in your home. You can also dim and brighten lamps.

The Controllers tell the Modules what to do. Command signals are sent over your existing house wiring to the Module or Modules of your choice. The Modules respond to the command signals. The Lamp Module turns on and off, dims and brightens any incandescent lamp up to 300 watts. The Appliance Module turns an appliance like a TV, fan or stereo on and off. It can also be used for a lamp - but can't dim it. The Wall Switch Module turns on and off, dims and brightens any incandescent light up to 500 watts, which is normally operated by a wall switch.

There are many different types of Controllers available:

Timers - for automatic control

The CR512 is a clock radio, a timer and a controller all in one. It controls 8 lights and appliances instantly and 4 automatically.

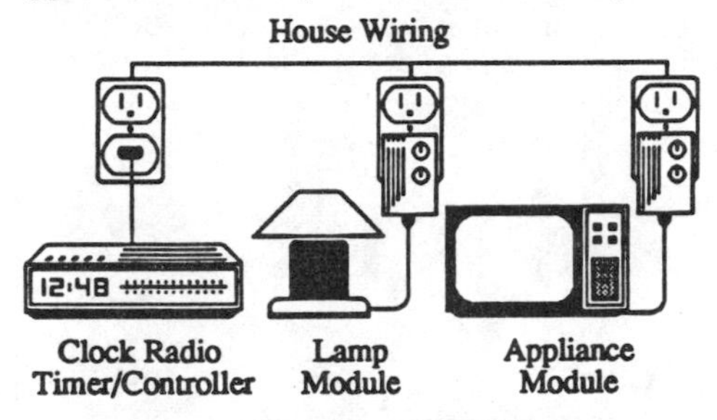

The MT522 is a timer, a controller and an alarm clock all in one. It controls 8 lights and appliances instantly and 4 automatically.

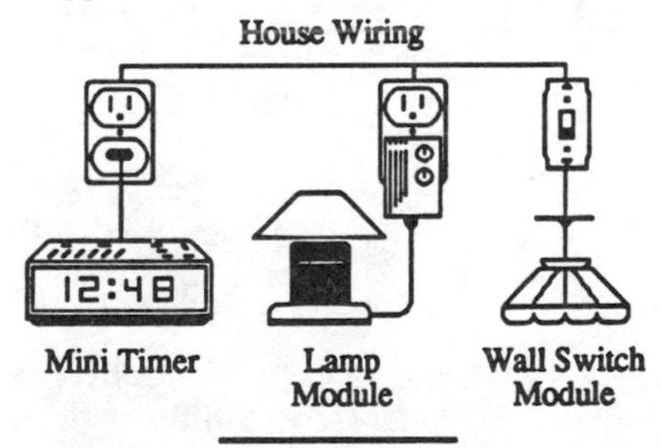

The CP290 is set up from a computer and then disconnected. It controls 256 X-10 Modules with 128 timed events.

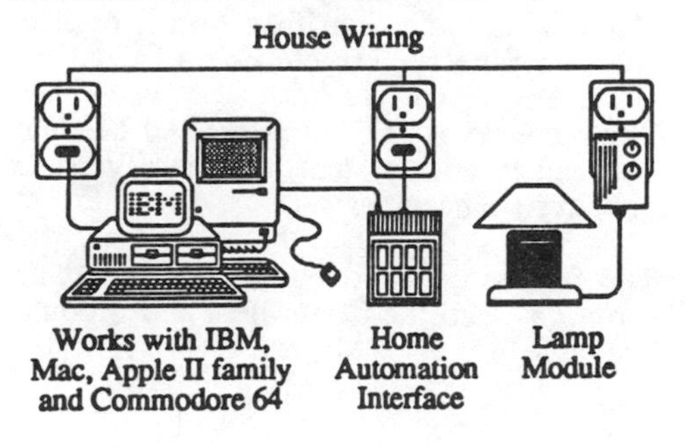

Manual Controllers - for remote control

The RC5000 set includes a Wireless Controller which controls a light or appliance plugged into the Transceiver Appliance Module... from inside or outside your home. It also controls 8 additional X-10 Modules.

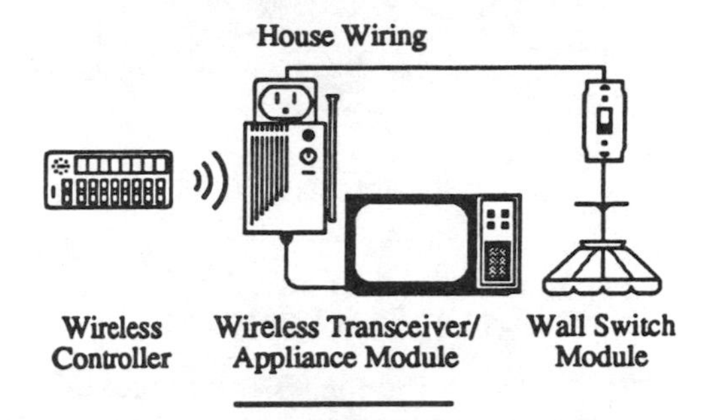

The Mini Controller controls up to 8 X-10 Modules at the touch of a button and dims lights too.

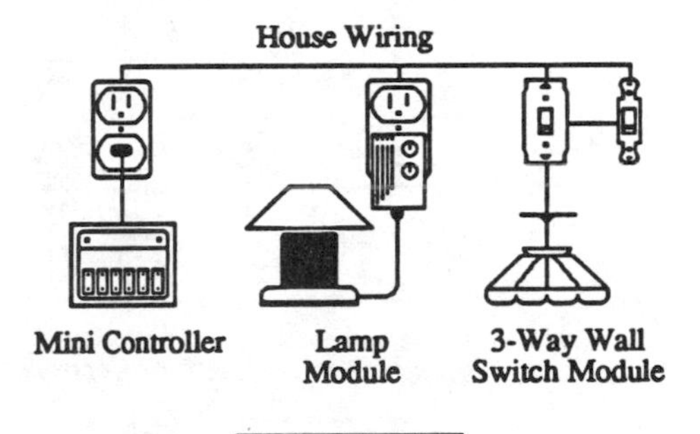

The Maxi Controller controls up to 16 X-10 Modules and can dim and brighten lights in groups.

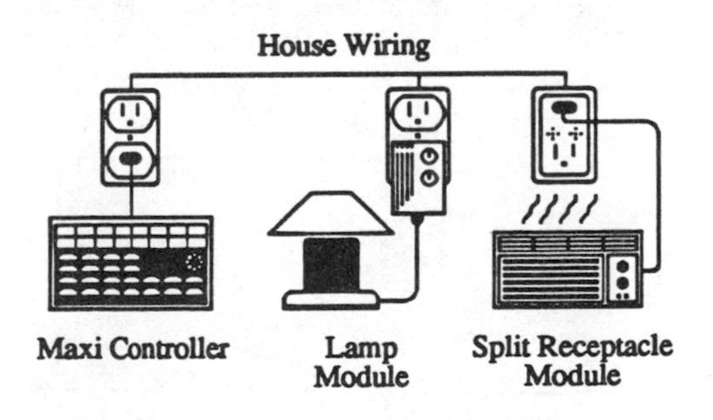

SUNDOWNER™ Turns lights on automatically when the sun goes down and turns them off again at dawn.

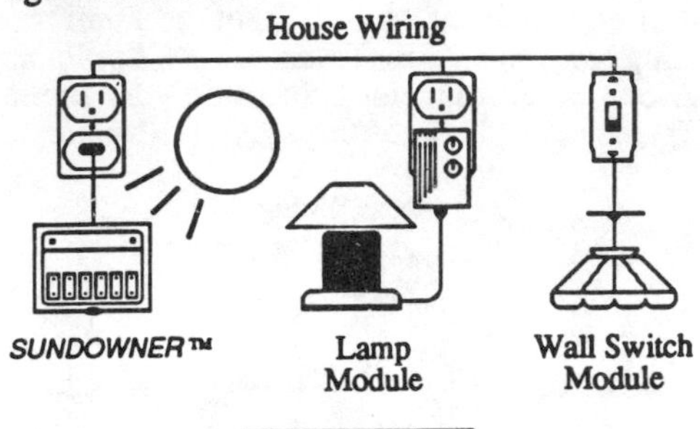

Telephone Responder and Thermostat Set-Back Controller

The Telephone Responder can control up to 8 lights and appliances from any phone in the world! When used with the Thermostat Set-Back Controller it can also control your central heating and air conditioning.

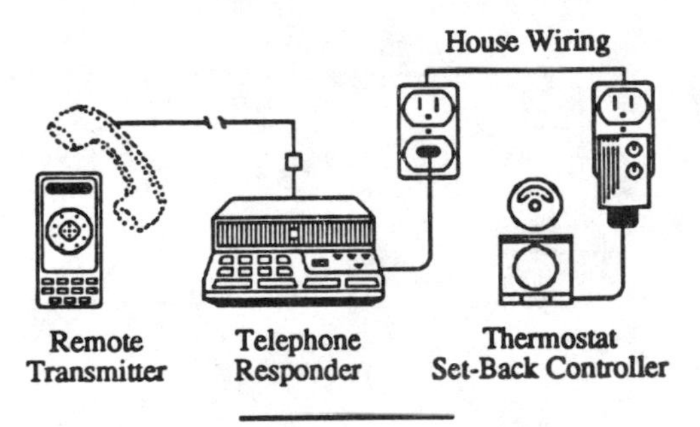

Burglar Alarm Interface

The Burglar Alarm Interface connects to the output of an existing alarm panel and flashes all the lights connected to Lamp Modules and Wall Switch Modules when the alarm trips.

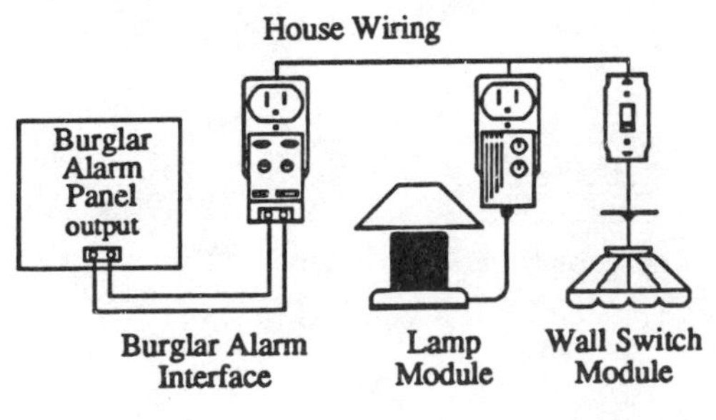

For the Ultimate in Security

The *PROTECTOR PLUS*™ Supervised Home Security System is a wireless security system which is compatible with all X-10 products. The Door/Window Sensors send wireless signals to the Base Receiver to set off its alarm. The Base Receiver also sends signals over existing house wiring to flash lights connected to X-10 Modules. It can also trip the *POWERHORN*™ Remote Power Line Siren. You can add the *SIXTEEN PLUS*™ Remote Control which lets you turn on and off, or brighten and dim up to 16 lights from anywhere inside or outside your home.

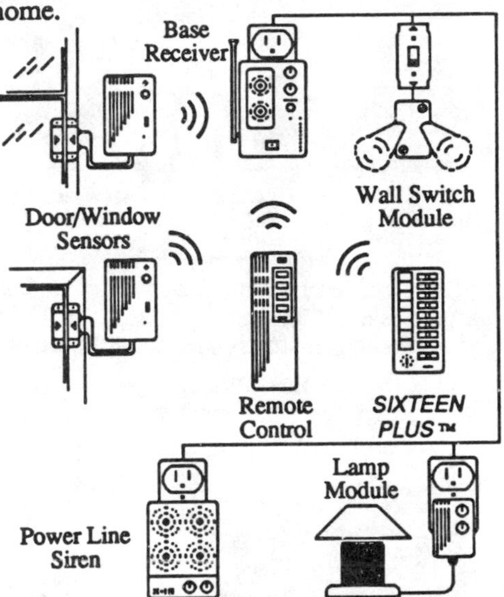

All *X-10 POWERHOUSE* Controllers and Modules are totally compatible with each other so you can use any of the Controllers with any of the Modules. You can turn a light on from a Mini Controller and have it set to go off automatically from the Clock Radio Timer. You can then turn it back on from the Wireless Remote Control.

The more types of Controllers and Modules you add to your system, the more versatile your system becomes.

The following section gives a detailed description of all the Controllers and Modules available.

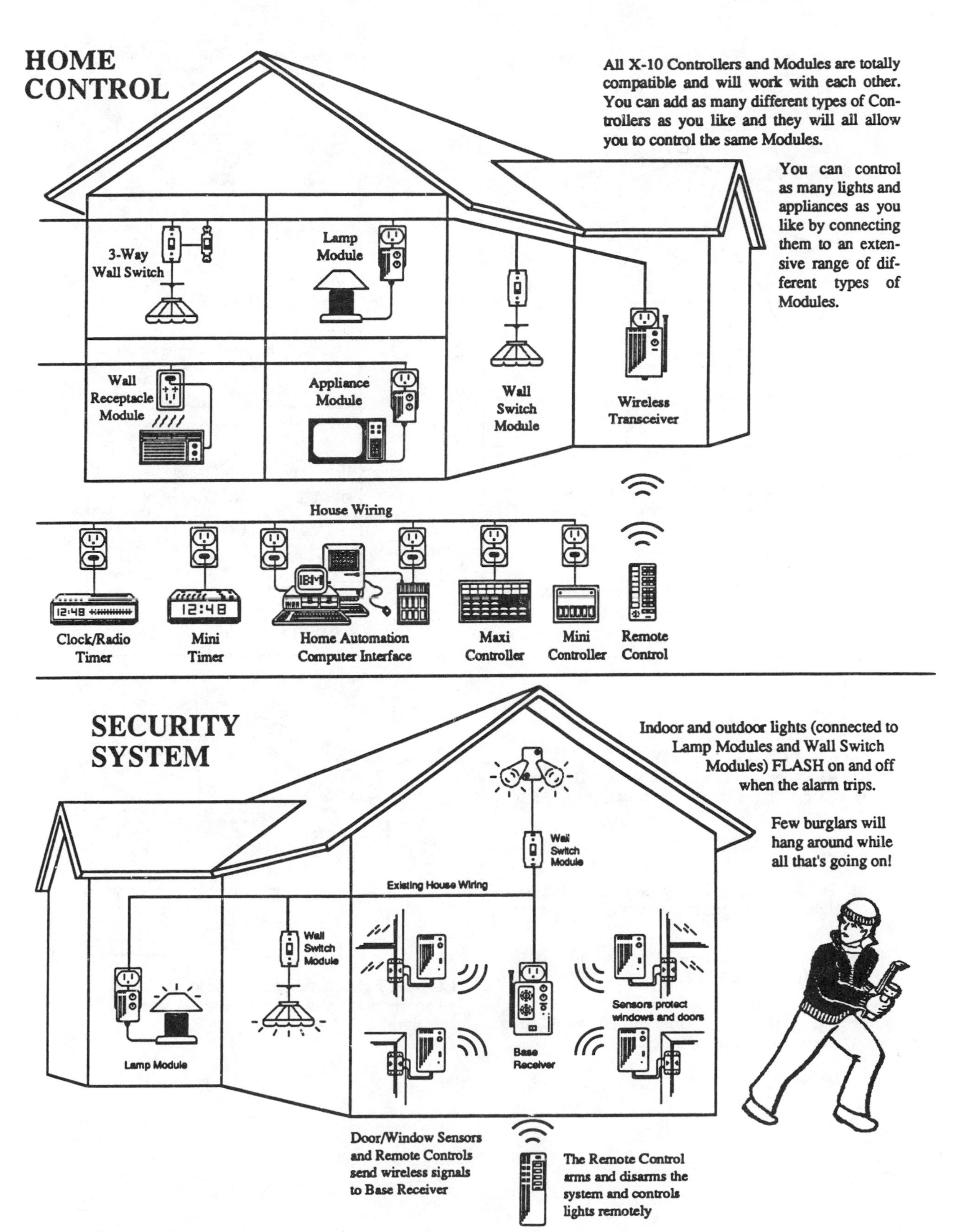
HOME CONTROL
All X-10 Controllers and Modules are totally compatible and will work with each other. You can add as many different types of Controllers as you like and they will all allow you to control the same Modules.
You can control as many lights and appliances as you like by connecting them to an extensive range of different types of Modules.
3-Way Wall Switch
Lamp Module
Wall Receptacle Module
Appliance Module
Wall Switch Module
Wireless Transceiver
House Wiring
Clock/Radio Timer
Mini Timer
Home Automation Computer Interface
Maxi Controller
Mini Controller
Remote Control
SECURITY SYSTEM
Indoor and outdoor lights (connected to Lamp Modules and Wall Switch Modules) FLASH on and off when the alarm trips.
Few burglars will hang around while all that's going on!
Wall Switch Module
Existing House Wiring
Wall Switch Module
Lamp Module
Sensors protect windows and doors
Base Receiver
Door/Window Sensors and Remote Controls send wireless signals to Base Receiver
The Remote Control arms and disarms the system and controls lights remotely

Dual Floodlight Outdoor Motion Detector with remote light control

Model No. PR511

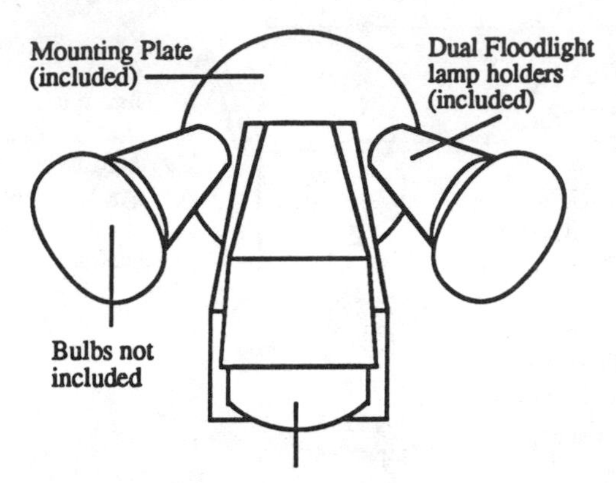

Motion Detector/Light Sensor:

① Turns on the connected floodlights either when motion is detected or when it gets dark.
② Turns on up to four X-10 Modules when motion is detected.
③ Turns on up to four X-10 Modules at dusk and turns them off again at dawn.

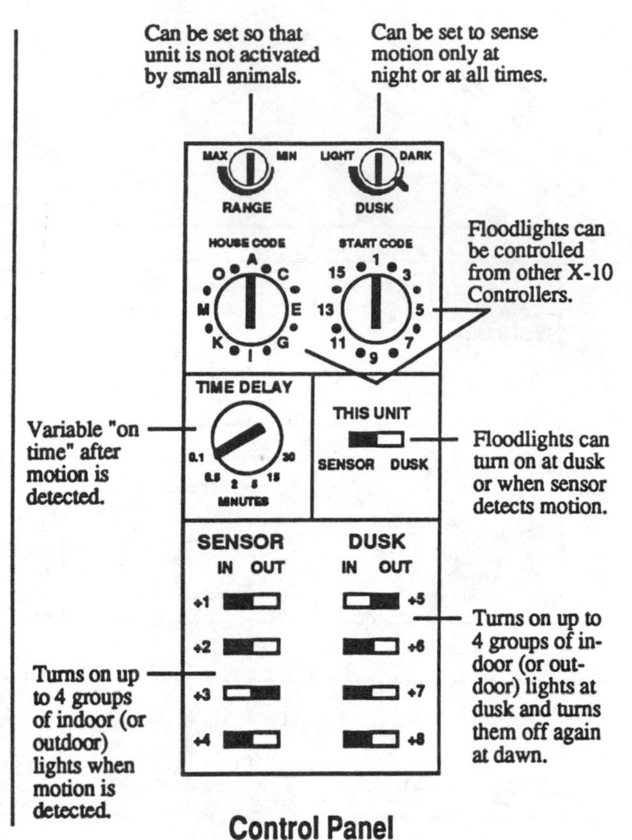

Control Panel

APPLICATIONS

- Replaces an existing outdoor floodlight fixture or can be mounted in a new location.
- Turns floodlights on at dusk and off at dawn, or when motion is detected.
- Turns on up to four groups of additional lights or appliances when motion is detected.
- Turns on up to four groups of additional lights at dusk and turns them off again at dawn.
- Floodlights can also be turned on and off from any X-10 Controller.

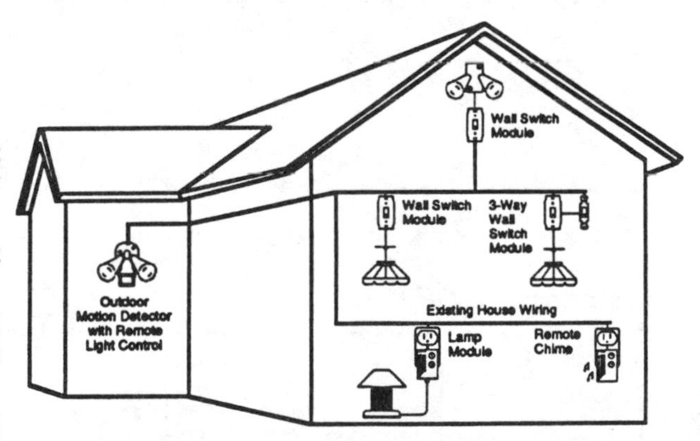

Controls lights inside and outside the home when motion is dectected and/or when it gets dark.

Mini Timer

Model No. MT522

The MT522 is a Mini Controller, a Timer and an Alarm Clock all in one.

You can set the Mini Timer to turn on and off up to four X-10 Modules twice a day. Each Module can be set to turn on or off ONCE only, EVERYDAY at the same time or in the SECURITY mode. The Security mode will vary the on or off time you set to be different each day (within the hour you set) to give your home a "lived-in" appearance. Rocker key number one sets the time for Modules set to Unit Code 1 and also sets the time for the Mini Timer's internal wake-up beeper.

You can turn on and off, up to eight Modules instantly by pressing any of the four on/off rocker keys after first setting the 1-4/ 5-8 slide switch to the desired position. You can also dim and brighten lights connected to Lamp Modules and Wall Switch Modules at the touch of a button.

If you press the Delay button after you turn a Module ON, the Mini Timer will automatically turn it OFF 15 minutes later, two presses of the Delay key gives 30 minutes of ON time, 3 presses gives 45 minutes and so on. You can also use the Delay key to turn a Module OFF for 15, 30, 45 minutes etc.

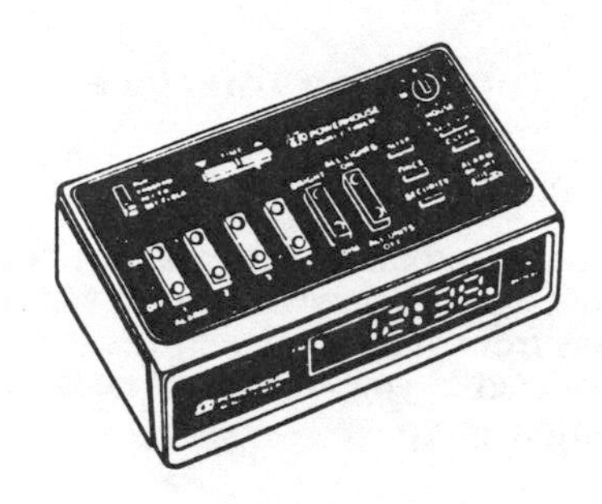

There is an All Lights On/All Units Off rocker key to instantly turn on all lights connected to Lamp Modules and Wall Switch Modules or turn off everything in your system.

The battery back-up will protect the time and programmed events during a power outage of up to 48 hours. Uses a 9V Alkaline battery (not included).

APPLICATIONS

- Set your outside lights as well as inside lights and appliances to go on and off at pre-set times to give your home that lived-in look while you're away.
- Use the "Security" button to automatically vary your programmed time each day so your home doesn't look like it's controlled by a timer.
- Keep the Mini Timer by your bedside and its built-in alarm will wake you up on weekdays. It has a snooze feature so you can sleep a while longer or you can turn the alarm off on weekends.
- Instantly control up to eight Modules, e.g. turn on your TV, dim your bedside light.
- Turn off your child's radio or night light from your bedside.
- Turn on ALL lights connected to Lamp and Wall Switch Modules with the touch of a button if you hear a strange noise at night.
- Turn off everything in the system with one button when you go to bed.

Home Automation Interface

Model No. CP290

The Home Automation Interface is best described as a very powerful timer which you program from your computer. It can then be disconnected if you wish, and **does not tie the computer up.**

It is an RS-232 Interface and works with most major brands of computer. It is supplied complete with the appropriate software and connecting cable for either IBM, Macintosh Apple IIe/IIc or Commodore 64/128. If you own another kind of computer, you can write your own software with the aid of the programming guide supplied, but for most people no knowledge of programming is required to set up and operate the Interface.

The Interface can address all 256 X-10 codes (16 Unit Codes X 16 Housecodes). The actual number of Modules which can be controlled is software dependant (256 for IBM and Macintosh, 72 for Apple IIe/IIc, 95 for Commodore 64/128). You can program different times for different days and you can even program the brightness level of lights.

The Interface can store 128 "Timed Events".

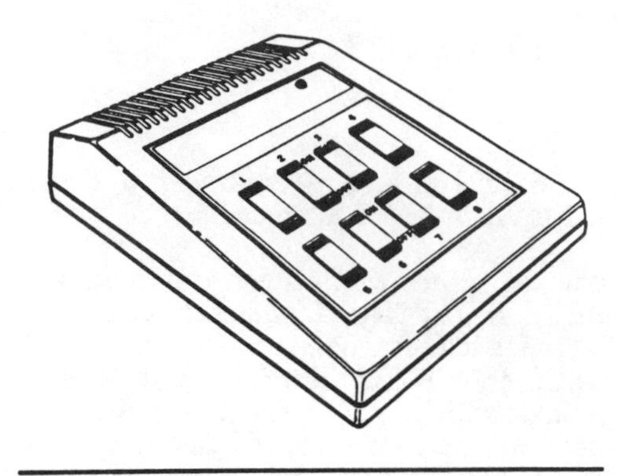

A "Timed Event" can be up to 16 devices on the same Housecode programmed to go on or off at a particular time on a particular day or days.

For example: Modules A1, A4, A7 and A16 programmed to go on at 7:30 p.m. at 40% brightness on Monday, Wednesday and Friday is just one event.

The Interface has its own real time clock and has battery back-up to protect the time and program for up to 100 hours using a 9V alkaline battery (not supplied). There are 8 rocker keys on the Interface which provide manual override for your first 8 Modules whether or not the Interface is connected to the computer.

APPLICATIONS

- Automate your home to run automatically, to give it a "lived in" appearance to deter intruders.
- Set the Interface to turn on outdoor security lights at night and turn them off again the next morning.
- Automatically set back your central heating or air conditioning.
- Fall asleep to the TV and wake up to stereo sound.
- Have your coffee ready when you get up.
- If you don't own a computer but have access to one. You can set up the Interface on a PC at work or on a friend's computer then install a battery and take the interface home. The battery will protect the memory until you get home and plug the interface into a 120V outlet.
- An ideal tool for hackers. Use the interface's real time clock to add a clock to your system for use with other programs. Take in inputs to your computer (through joystick ports etc.) then have your program make decisions based on these inputs and download instructions to the Interface. You could turn on outside lights when a photocell tells your computer that it's dark outside, or turn off a water heater when your computer detects that power usage has exceeded a predetermined limit.

Software & cable for IBM and compatibles

For use with the CP290 Home Automation Interface. This program is text driven and although it supports color, does not require a graphics card. The user enters timed events in cells, which represent the X-10 code, type of light or appliance and location in the house. The program will address all 256 codes available and will allow programming of different events for different days. You can save schedules to disk (for Summer/Winter/Holiday schedules etc.) and can print a hard copy of your schedules.

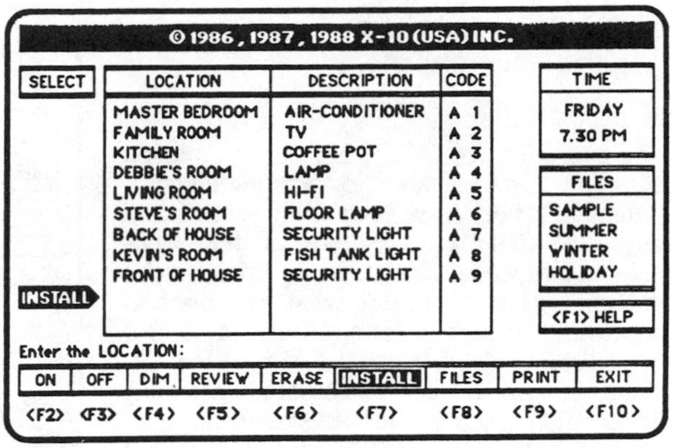

Selections are made with either the cursor keys or the function keys

Cursor control, or Function keys if you prefer

The IBM compatible software is text driven and does not require a color monitor or graphics card. It will run on any 100% IBM compatible machine. Cells are selected with the up and down cursor keys and then operations, such as ON, OFF, INSTALL, REVIEW, are selected by using the left and right cursor keys. Selections are then confirmed with the RETURN key. You can also use the Function keys to select these options.

Install what you want, where you want it

Lamps and appliances to be controlled are installed by selecting INSTALL with the cursor keys. You then select the cell in which you want to install something and simply type in the name, location and X-10 code setting for the light or appliance to be installed. Up to 256 differently coded X-10 modules can be installed. You can scroll through the screen to view any group of cells, 15 cells (lights or appliances) are in view on the screen at any one time.

Step through the menus with ease

The options at the bottom of the screen, <F1>, <F2>, etc. change as you step through the program; i.e. after selecting ON, the bottom line changes to: NOW, TODAY, TOMORROW, WEEKENDS, WEEKDAYS, EVERYDAY, SPECIFIC DAYS. You then use the left and right cursor keys to select when you want the event to occur and enter your selection with the RETURN key. You are then asked to enter the HOURS and MINUTES for any timed event.

Advanced features

You can group up to 16 different lights or appliances (on the same Housecode) together as one "EVENT", and you can program up to 128 EVENTS. These EVENTS can be saved to disk as a SCHEDULE and you can save different schedules, for different times of the year for example. You can print a hard copy of your schedules. There is even a handy FREEZE feature which suspends an event until told to UNFREEZE it. This is useful if you are entertaining guests, for example, and want to suspend the usual program of events for that evening.

Software & cable for Macintosh computers

For use with the CP290 Home Automation Interface. The software for Macintosh computers lets you draw your own background with MacPaint or use one of the backgrounds supplied. You install icons on the background to represent what you want to control. You then point and click to control up to **256** differently coded X-10 Modules. There is also a Desk Accessory which lets you control your home while working with another software application. The software is fully compatible with the Mac SE, Mac II and Multifinder.

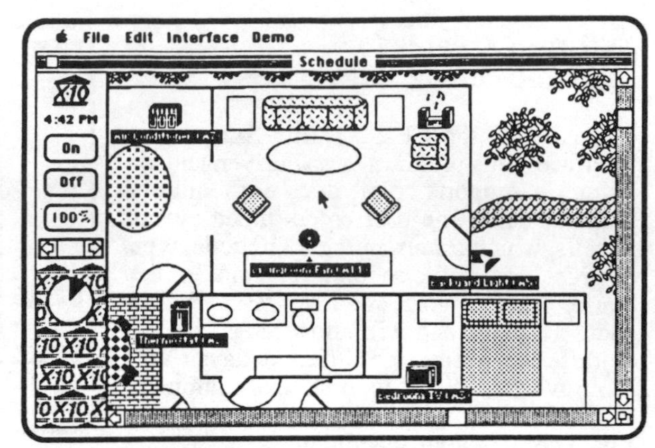

You can install icons on one of the backgrounds supplied or draw your own. You can even customize the icons.

The Module Map (right part of the screen) is a floor plan of a typical home. You can change this with MacPaint to look like your home. You can select icons (pictures of lights and appliances), then set the X-10 code for the Module represented by the icon to any Housecode from A to P and any Unit Code from 1 to 16 (256 combinations). You can name the icon anything you wish, such as "Family Room TV", you can even draw your own icons. You can then scroll through the Module Map and place the icons anywhere you wish.

The Control area (left part of the screen) lets you select what you want to happen to the Module(s) you selected in the Module Map. You can click the ON-OFF button or select an intensity (for lamps). The Interface will then instantly turn on or off the selected Module(s).

To program an event, double click on an icon to open its Events Window. Then click the switch for ON or OFF, or set the intensity, advance to the time you wish, click Today, Tomorrow or Weekly and then click on the day(s) you would like to program. If you click on the security button, this will randomly vary your programmed event to be within 30 minutes of the time you set (to give your home a "lived-in" appearance).

The "Events Window" for a lamp or appliance appears when you double click on its icon.

You can select New Event and add events to your "Event List". You can store up to 128 events in the CP290 Interface and you can group up to 16 Modules on the same Housecode to be part of one event. There is a Desk Accessory which lets you control lights and appliances instantly while working with another application.

Cable for the Mac 512k supplied. Cable adaptor required for Mac Plus, Mac SE, and Mac II etc.

Software & cable for Apple IIe/IIc

For use with the CP290 Home Automation Interface. The software displays a colorful graphical representation of the rooms in a house and allows you to install ICONS (graphical pictures of lights and appliances) in the rooms. Then using a Joystick, Mouse or the Keyboard cursor keys you select the icons and set the times you want the lights or appliances to turn on or off. You can choose to turn them on Now, or at specific times later Today, Tomorrow, Everyday or on Specific Days. You can install up to 72 ICONS, and therefore, can control up to 72 lights and appliances. You can save the programmed events onto the disk which lets you have different schedules for different times of the year.

For programmers:

The disk also contains a Utility Program which (if you desire) can help you to write your own software in BASIC. The package includes both a "Din to Din" cable for the IIc and a "Din to DB-25" cable for the IIe.

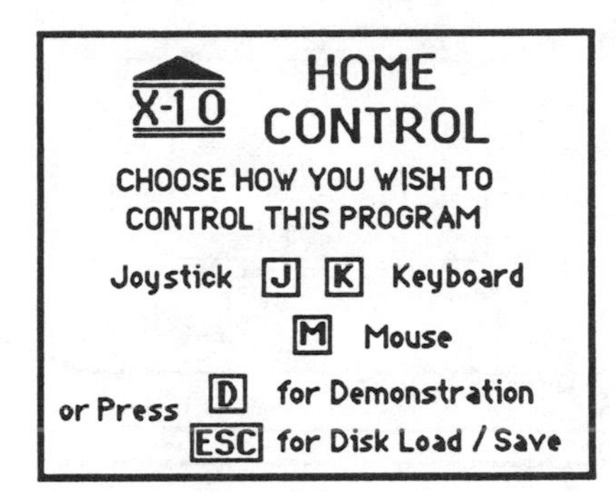

The Introductory screen lets you choose how you want to use the program - with a Joystick, Mouse or with the Keyboard cursor keys. You can also select the Demo mode or press the Escape key to save your schedules to disk.

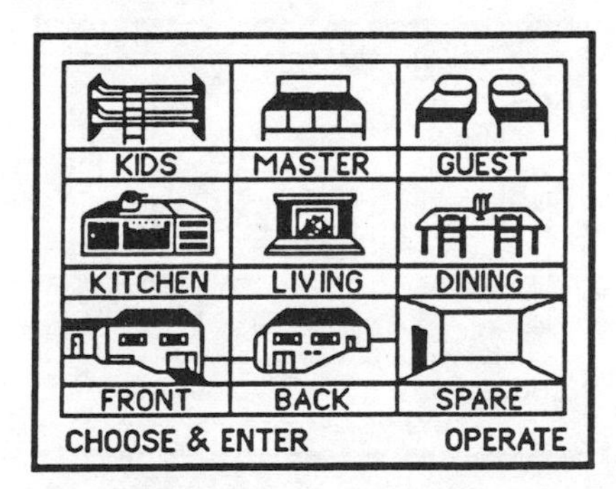

The Multi-room screen allows you to choose a room with a Joystick, Mouse or the Keyboard cursor keys. You then enter the room by pressing the button on the Joystick or Mouse or the Return key on the keyboard.

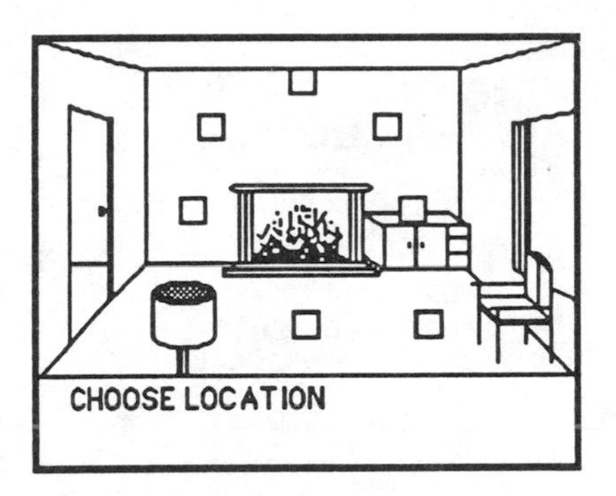

In the **INSTALL** mode you can place up to 8 ICONS in each room. You can choose different ICONS for various lights and appliances. You also set the master clock in the INSTALL mode.

In the **OPERATE** mode you can select an ICON in any room and then select what you want to program for that light or appliance. Your commands are automatically stored in the CP290 Home Automation Interface.

Software & cable for Commodore 64/128

For use with the CP290 Home Automation Interface. The software displays a colorful graphical representation of the rooms in a house and allows you to install ICONS (graphical pictures of lights and appliances) in the rooms. Then using a Joystick or the Keyboard cursor keys you select the icons and set the times you want the lights or appliances to turn on or off. You can choose to turn them on Now, or at specific times later Today, Tomorrow, Everyday or on Specific Days. You can install up to 95 ICONS, and therefore, can control up to 95 lights and appliances. You can save the programmed events onto the disk which lets you have different schedules for different times of the year.

For programmers:

The disk also contains a Utility Program which (if you desire) can help you to write your own software in BASIC. The package includes a cable for connecting the CP290 to the "user port".

The Introductory screen lets you choose how you want to use the program - with a Joystick or with the Keyboard cursor keys. You can also select the Demo mode for a demonstration.

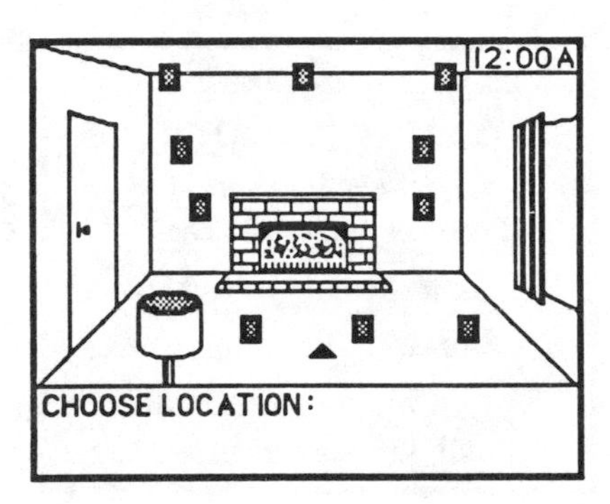

In the **INSTALL** mode you can place up to 11 ICONS in each room. You can choose different ICONS for various lights and appliances. You also set the master clock in the INSTALL mode.

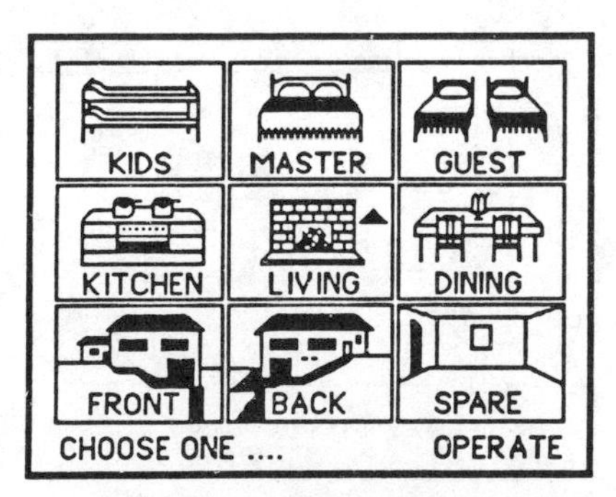

The Multi-room screen allows you to choose a room with a Joystick or the Keyboard cursor keys. You then enter the room by pressing the button on the Joystick or the Return key on the computer keyboard.

In the **OPERATE** mode you can select an ICON in any room and then select what you want to program for that light or appliance. Your commands are automatically stored in the CP290 Home Automation Interface.

Wireles Remote Control System

Model No. RC5000

The RC5000 is a complete and ready to use system. It consists of an eight unit, four function Wireless Remote Control (RT504) which transmits radio frequency (RF) signals to turn on or off a light or appliance plugged into the Transceiver/Appliance Module (RR501). So you can remotely turn on a light or appliance from inside or outside your home.

The Transceiver/Appliance Module however is much more than just an Appliance Module.

1. It can be turned on and off from the Wireless Remote Control.
2. It can be turned on and off from any other X-10 Controller, such as the Mini Timer.
3. It re-transmits, over your house wiring, any signals it receives from the Wireless Remote Control. This lets you control up to eight additional X-10 Modules from the Wireless Remote Control.

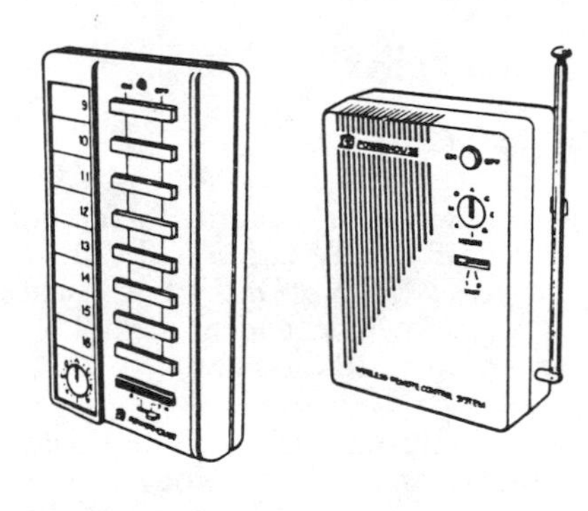

(one of the other Modules is set to the same Unit Code as the Transceiver/Appliance Module).

You can also dim and brighten lights connected to X-10 Lamp Modules and Wall Switch Modules from the Wireless Remote Control.

The Wireless Remote Control and the Transceiver/Appliance Module have a "bank switch" which allows you to have 2 systems in use in the same house. The first set can control Modules set to 1 thru 8 and the second set can control Modules set to 9 thru 16.

APPLICATIONS

- Control lights and appliances from anywhere inside or outside your house - in the yard or from your car. Imagine being able to turn on the lights in your house as you enter your driveway. You will never have to enter a dark house again.
- Add up to eight additional X-10 Modules to your system.
- Add another Transceiver/Appliance Module to control up to sixteen Modules. You set one Transceiver to control Modules 1-8 and the other to control Modules 9-16.
- Attach the Wireless Remote Control to the wall inside a doorway (using Velcro™) so you can turn on lights as you enter the room.
- Add additional Wireless Remote Controls, leave one in the house and have one in your car.
- Give one to an elderly parent or someone who is disabled so they can flash all the lights in the system in the case of an emergency.
- Keep one by the bedside so you can turn on lights if you hear a strange noise in the middle of the night.
- A disabled person could use the Wireless Remote Control from a wheelchair. The severely disabled could use a mouthstick to operate the buttons on the Wireless Remote Control which could be easily mounted on the wheelchair's armrest.

Maxi Controller

Model No. SC503

The Maxi Controller performs 6 functions: ON, OFF, BRIGHT, DIM, ALL LIGHTS ON and ALL UNITS OFF. It has sixteen number keys for independent control of up to 16 groups of X-10 Modules.

The Maxi Controller has "two key action". This means that to turn a Module on, you first press and release the number button corresponding to the Unit Code set on the Module you want to turn on. You then press and release the ON button. This two key action provides an added feature. It allows "group control" and group dimming/brightening.

Group Control:

Press and release buttons 1, 2 and 3, etc. in succession. Then press the ON button. Modules set to 1, 2 and 3, etc. will all turn on at the same time.

Group Dimming:

Most brands of dimmer allow only individual manual control. The Maxi Controller not only allows you to *remotely* dim lights connected to X-10 Lamp Modules and Wall Switch Modules but also allows you to dim and brighten them in groups! Press buttons 1, 2 and 3, etc. in succession. Then press the dim button to dim them all at the same time. The same procedure is used to brighten lights. X-10 dimmers are the only dimmers that have this feature.

APPLICATIONS

- Lets you remotely control up to sixteen X-10 Modules from anywhere in your home.
- Dim "groups" of lights connected to Lamp Modules and Wall Switch Modules at the same time, for that special "mood" setting.
- Keep the Maxi Controller by your bedside and turn off the TV without getting out of bed.
- Dim and brighten your bedside lamp.
- Turn off your child's radio or night light from your bedroom.
- Turn on ALL lights connected to Lamp Modules and Wall Switch Modules with the touch of a button if you hear a suspicious noise at night.
- Turn off everything in the system with one button when you go to bed.
- Turn your central heating or air conditioning down using the Thermostat Set-Back Controller.

Mini Controller

Model No. MC460

The Mini Controller performs 6 functions: ON, OFF, BRIGHT, DIM, ALL LIGHTS ON and ALL UNITS OFF. It can be used from your bedside for example or from any convenient location in your home. It's so inexpensive that you will want to have one in every room!

It has four "rocker" keys for one touch ease of use. These keys can control Modules set to Unit Codes 1 thru 4 or 5 thru 8 by sliding the 1-4/5-8 selector switch to the desired position.

The Mini Controller can dim and brighten lights connected to Lamp Modules and Wall Switch Modules.

APPLICATIONS

- Lets you remotely control up to eight X-10 Modules from anywhere in your home.
- Keep it by your bedside and turn off the TV without getting out of bed.
- Turn off your child's radio or night light from your bedroom.
- Dim and brighten your bedside lamp.
- Turn on ALL lights connected to Lamp Modules and Wall Switch Modules with the touch of a button if you hear a suspicious noise at night.
- Turn off everything in the system with one button when you go to bed.
- Turn your central heating or air conditioning down using the Thermostat Set-Back Controller.

SUNDOWNER™

Model No. SD533

Its built-in photocell turns on up to four lights when the sun goes down and turns them off again at dawn - automatically, with no programming!

There are four slide switches on the underside of the *SUNDOWNER* which enable you to select which Modules you want it to control in the dusk-to-dawn mode.

There is a sensitivity adjustment which allows you to set the light level at which the *SUNDOWNER* detects dawn.

The *SUNDOWNER* also has all the features of the Mini Controller. It has four "rocker" keys for one touch ease of use. These keys can control Modules set to Unit Codes 1 through 4 or 5 through 8 by sliding the 1-4/5-8 selector switch to the desired position.

The *SUNDOWNER* performs 6 functions: ON, OFF, BRIGHT, DIM, ALL LIGHTS ON and ALL UNITS OFF. It can be used from your bedside for example or from any convenient location in your home.

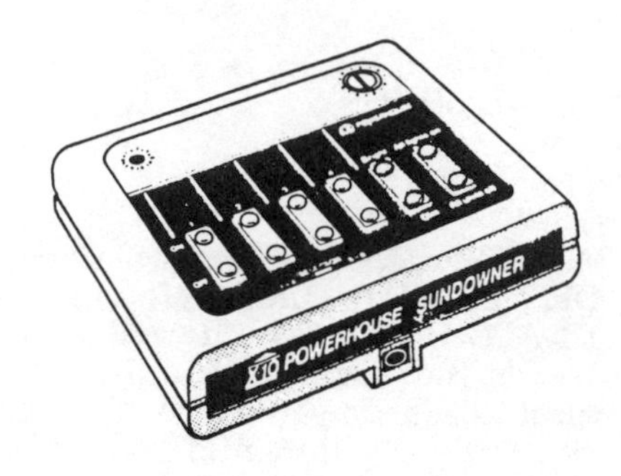

It can turn lights and appliances on and off by remote control and can dim and brighten lights connected to Lamp Modules and Wall Switch Modules, Appliance Modules do not respond to dim and bright commands.

APPLICATIONS

- Makes your home look lived-in by turning lights on automatically at dusk... without having to program any timers.
- Saves energy by turning lights off automatically at dawn.
- Lets you remotely control up to eight X-10 Modules from anywhere in your home.
- Keep it by your bedside and turn off the TV without getting out of bed.
- Turn off your child's radio or night light from your bedroom.
- Dim and brighten your bedside lamp.
- Turn on ALL lights connected to Lamp Modules and Wall Switch Modules with the touch of a button if you hear a suspicious noise at night.
- Turn off everything in the system with one button when you go to bed.

Telephone Responder

Model No. TR551

The Telephone Responder (TR551) is a Controller which plugs into a standard 120 volt outlet and a standard modular telephone jack. When you call home, **from any touch-tone phone in the world**, the TR551 answers the phone, just like an answering machine. It takes about 30-35 seconds before it answers and then you hear three "beeps." You then enter your commands from the telephone's touch-tone keypad to turn on or off any of up to ten Modules. Digits 1-9 represent Modules with Unit Codes 1-9 and digit 0 on the touch-tone keypad represents Module number 10. You enter the desired number followed by * for ON or # for off.

You can set a "security code" of up to three digits to prevent unauthorized use. There are 1000 security codes to choose from.

You can set the TR551 to flash selected lights around the house when the phone rings. Just set the desired Modules to the same number as the third digit of the security code you selected. This is a useful feature for the hearing impaired.

You can use the TR551 even if you already own an answering machine, just set the slide switch on the TR551 to the answering machine position. Most answering machines

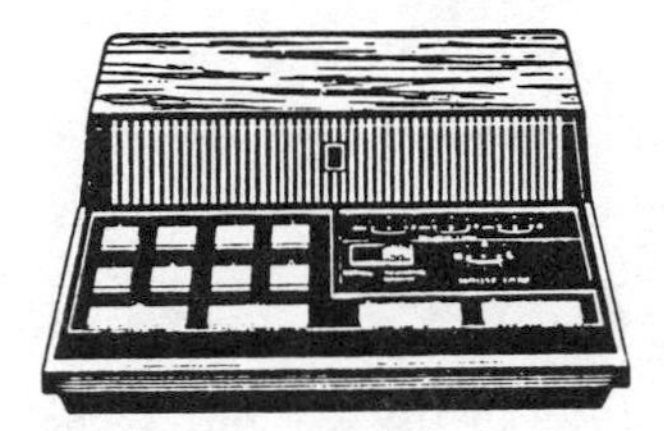

answer the phone after about three rings, the TR551 answers 30-35 seconds after the first ring even though the ringing stopped when your answering machine picked up. If you don't own an answering machine you set the slide switch on the TR551 to the "normal" position. This prevents the TR551 from answering if the line has already been picked up by someone, so you don't have explain to your callers what the three beeps are.

When you are at home you can use the TR551 as a Manual Controller to control up to eight X-10 Modules.

The TR551 also has an "All Lights On" button to turn on ALL Lamp Modules and Wall Switch Modules at the touch of a button and an "All Units Off" button to turn off all Modules which are set to the same Housecode as the TR551.

APPLICATIONS

- Call your beach house and turn on the air conditioning before you leave for your vacation.
- Call your ski cottage to turn on the heating before you leave for a weekend of skiing.
- Purchase two TR551's so that you can control lights and appliances in your "main" home while you are away at your second home - to make it look like you are still at home.
- If you have an irregular schedule, use the Telephone Responder to call home and turn on the lights as well as heating, air conditioning etc. before you leave to go home.
- Call home and turn off your sprinkler system if you notice that it has started raining.
- Flash selected lights around the house when the phone rings - useful for the hearing impaired.

POWERFLASH™
Burglar Alarm Interface

Model No. PF284/BA284

Connects to an existing alarm system and turns on lights when the alarm trips. It has 3 modes of operation:

In Mode 1 the PF284 will turn ON all Lamp Modules and Wall Switch Modules set to its Housecode and will also turn ON any other Modules set to its Unit Code, a stereo connected to an Appliance Module for example. All Lamp Modules and Wall Switch Modules are left in the ON state when the alarm is de-activated but the Modules set to the same Unit Code as the PF284 will be turned OFF.

In Mode 2 the PF284 will FLASH all lights connected to Lamp Modules or Wall Switch Modules. All Lamp and Wall Switch Modules will be left in the ON state when the alarm is de-activated but **Appliance Modules** set to the same Housecode as the PF284 will be turned OFF.

In Mode 3 the PF284 will turn ON all Lamp Modules and Wall Switch Modules set to the same Housecode as the interface. All Lamp Modules and Wall Switch Modules will be left in the ON state when the alarm is de-activated. Mode 3 is the same as mode 1 except that the interface does not send the code to turn on Modules set to the Unit Code setting on the interface.

The interface can be set to be triggered by either a low voltage input (6-18 volts AC, DC or audio) or a dry contact closure. You could also trigger it from an alarm which has an outlet for switching a 120V lamp by using this outlet to control a 120V relay, then use the contacts of the relay to trigger the interface (in the dry contact mode). **DO NOT CONNECT 120VAC DIRECTLY TO THE INPUT OF THE INTERFACE.**

The PF284 has a test button which allows you to activate it whether or not it is connected to an alarm system. The ALL OFF button allows you to turn off anything which has been turned on by the interface.

APPLICATIONS

- Flash all the lights in your home when your existing burglar alarm is tripped.
- Blast your stereo when your alarm trips.
- Connect the PF284 directly to a magnetic switch to turn on lights when a door or window opens, i.e. you don't have to own a burglar alarm system.
- Trigger it from sensors such as a photo cell to turn on lights when it gets dark, from a motion detector to turn on lights when someone enters a room, from a microphone to turn on lights or appliances when a sound from an intruder is detected, from a moisture sensor to sound an alarm if your basement is flooded, from a garage door opener to turn on lights when you open your garage door, etc.

Some of these applications may require some additional components such as relays, etc. to interface to the PF284 but virtually anything which gives, or can be made to give, either a contact closure or a low voltage output between 6 and 18 volts AC, DC or audio can be used to trigger the PF284.

Additional Applications for the PF284/BA284

The PF284 can be controlled from the "Big Switch" Motion Detector available from C & K Systems Inc. The system consists of 3 main components:

1. An Infrared sensor which detects motion and provides a contact closure.
2. An Automatic Light Control Interface (ALCI) which is similar to the PF284 and transmits signals onto the house wiring when triggered by a contact closure.
3. A Module which receives signals over the existing house wiring and turns on a light connected to its output.

In most cases the ALCI and the Wall Switch Module (or Lamp Module) can be anywhere in the house. However for reliable operation it is recommended that the following installation procedure be followed.

Installation

Follow these steps BEFORE choosing the location for the C & K sensor (model # 4060).

Wall Switch Module installation

1. Determine which ceiling or outside security light(s) are to be controlled.

2. Replace each wall switch which normally operates the light with the Wall Switch Module (model # 4064). Set the Housecode and the Unit Code on the Wall Switch Module(s) to A and 1 respectively.

3. Determine which breaker controls each wall switch. It is desirable (although not mandatory) to have all Wall Switch Modules connected to the same "phase" of the house wiring (see troubleshooting section for detailed explanation). If possible, reconnect the breakers to ensure that all wall switches are on the same phase. Note, breakers are wired with phase A and B connected alternately down each side of the panel, therefore any two adjacent breakers (one above the other) are *opposite* phases. If it is not possible to re-connect breakers to ensure that all modules are on the same phase, it is recommended that a coupling capacitor be fitted in the breaker panel (see troubleshooting section).

Lamp Module installation

1. Determine which lamp(s) are to be controlled.

2. Make sure that the on/off switch on the lamp to be controlled is in the ON position and plug the lamp into the Lamp Module (model # 4065). Set the Housecode and the Unit Code on the Lamp Module to A and 1 respectively.

3. Plug the lamp into the nearest convenient outlet *which is on the same phase* as the Wall Switch Modules. If this is not possible, re-connect the breaker for the desired outlet so that it is on the same phase or fit a coupling capacitor (see troubleshooting section). Repeat for other lamps to be controlled.

Automatic Light Control Interface (ALCI or PF284/BA284) installation

1. Determine the *approximate* location for the 4060 sensor. Locate the nearest convenient outlet *which is on the same phase* as the Wall Switch Modules and Lamp Modules that have been installed. Plug the Automatic Light Control Interface (model #4061), or PF284 into this outlet. Reconnect breakers to connect the chosen outlet to the correct phase if it isn't on the correct phase already or fit a coupling capacitor (see troubleshooting section).

2. Set the input switch on the Automatic Light Control Interface (ALCI) to the position marked V (position B on PF284).

3. Set the mode switch on the Automatic Light Control Interface (ALCI) to the po-

Light Control Interface (ALCI) to the position marked V (position 1 on PF284).

4. Set the Housecode and the Unit Code on the ALCI (or PF284) to A and 1.

5. Press (and hold) the Test button on the ALCI and check that **ALL** Modules *set to the same Housecode and Unit Code as the ALCI* turn on.

6. Release the Test button and check that **ALL** Modules *set to the same Housecode and Unit Code as the ALCI* turn off.

Refer to the troubleshooting section if you cannot get all modules to turn on and off.

NOTE.

ALL Wall Switch Modules and Lamp Modules set to the same Housecode as the ALCI will turn **on** when the Test button is pressed (regardless of Unit Code setting) but only those modules set to the same Unit Code as the ALCI will turn **off** when the Test button is released. Any modules (including any set to other Unit Codes) can be turned off by pressing "All Units Off" on the ALCI.

For the above set-up all Housecodes and Unit Codes were set to A and 1. You may if you wish, set the Housecode and Unit Code on the ALCI to any other letter and number but remember ALL modules to be controlled by the ALCI must be set to the same **Housecode.**

Any Lamp Modules and Wall Switch Modules which are NOT set to the same **Unit Code** as the ALCI will turn on when motion is detected but will stay on (even after the sensor times out) and must be turned off using the All Units Off button on the ALCI. If you want lights to turn on *and* off from the sensor, set the modules to the same Housecode *and* Unit Code as the ALCI.

If you set the Mode Switch on the ALCI to position 2, ALL Lamp Modules and Wall Switch Modules will FLASH when motion is detected and will stay on after the sensor times out.

If you set the Mode Switch on the ALCI to position 3, ALL Lamp Modules and Wall Switch Modules will turn on when motion is detected and will stay on after the sensor times out.

Sensor installation

1. Connect terminal 1 and 2 from the sensor to model # 1092 low voltage transformer.

2. Connect terminal 4 from the sensor to to + terminal of the ALCI.

3. Connect terminal 5 from the sensor to to - terminal of the ALCI.

4. Plug the transformer into the nearest AC outlet.

5. Refer to Installation Manual # 5051070 for more details regarding sensor set-up, adjustment, location etc.

When motion is detected, operation of lights should be the same as when pressing the Test button on the ALCI. When the sensor times out, operation of lights should be the same as when releasing the Test button on the ALCI.

Installation tips

When selecting an outlet for the ALCI try wherever possible **not** to plug it into the same outlet as a TV set or a stereo system. TV sets and stereos sometimes have a capacitor connected across their power cord and such a capacitor can attenuate the signals transmitted by the ALCI. In most cases the attenuation will not be noticeable but in marginal situations, where the ALCI and the Module are on opposite phases of the house wiring, this attenuation can make the difference between getting enough signal and not getting enough signal.

For more information on the **"The Big Switch"** motion detector, contact:

C & K Systems Inc.
2040 Fortune Drive, San Jose, CA 95131
Tel. (408) 434-1149 or 1-800-227-8065.

Thermostat Setback Controller

Model No. TH2807

The Temperature Regulator attaches to the wall under your existing thermostat and connects to the plug-in Power Supply. The Power Supply plugs into an Appliance Module or Wall Receptacle Module. When you turn on the Appliance Module or Wall Receptacle Module the Temperature Regulator supplies a small amount of local heat under your thermostat, fooling it into thinking that the room is hotter than it really is. Your heating is therefore set-back by the amount of heat which the Temperature Regulator supplies. You pre-set this for approximately 5, 10 or 15 degrees of setback. The reverse applies for central air conditioning. You simply set your thermostat for the maximum temperature that you would like the room to attain, then when you turn on the Appliance Module or Receptacle Module, the thermostat thinks the room is, say 15 degrees hotter, so it turns your air conditioning on. i.e. when the Appliance Module is OFF the room temperature will rise to the temperature setting on your thermostat and when the Appliance Module is ON, the room will be kept say 15 degrees cooler.

The Thermostat Set-back works with ANY kind of thermostat and does not require any

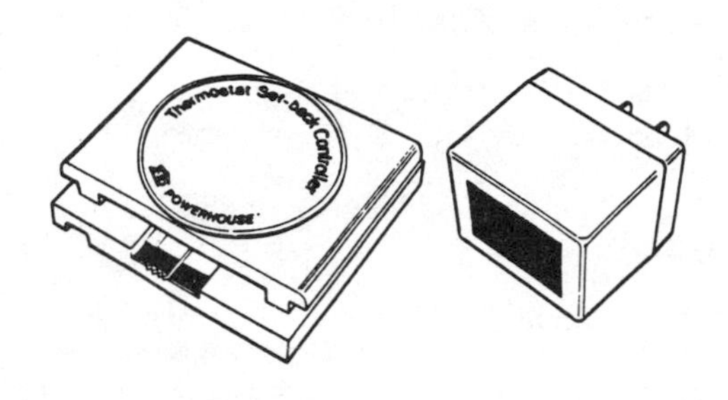

changes to the wiring of your existing thermostat. It will work with low voltage, high voltage or even pneumatic thermostats. You can automatically set-back heating or air conditioning using the Mini Timer, Clock Radio Timer or Home Automation Interface. You can also remotely turn up heating or air conditioning before you get home using the Telephone Responder. The ease of programming offered by X-10 Timers eliminates the problems often associated with programming set-back thermostats.

Just by setting your heating back by 15 degrees when you go to bed, you can conservatively save up to 20% a year on your heating bills so the system can pay for itself in a very short time.

APPLICATIONS

- Use it with any of X-10's Timers to set-back your heating or air conditioning while you're asleep or away from home.
- Use it with the Telephone Responder to turn up the heating in your winter home before you leave for a skiing vacation. or to turn up the air conditioning in your summer home.
- Use it with the Mini Controller to turn up the heating from your bedside if you feel a chill in the middle of the night.
- Use it with the RC5000 Radio Controller to turn up your air conditioning while you're watching TV without having to get up from your chair.

Appliance Module

Model No. AM486 - 2 pin polarized.
Model No. AM466 - 3 pin grounded.

The AM486 & AM466 are rated for three types of loads:

15A resistive. This is the equivalent of about 1800 Watts so it will control resistive type loads such as coffee pots, crock pots, toaster ovens, portable heaters etc.

Warning: extreme care should be taken when using the Appliance Module to control portable heaters as a fire could result if the heater were turned on by remote control while clothing was draped nearby.

1/3 HP for motor loads. Sufficient to handle the motor load in appliances such as portable fans and window air conditioners.

500 Watts for lamps. Although the Appliance Module is rated for 1800 Watts for resistive loads, it is limited to 500 Watts for incandescent lamps because of the "inrush" current when you turn on a cold lamp. This momentary inrush of current can be up to 6 times higher than the normal operating load.

The Appliance Module can be used for any kind of lamp including fluorescent lamps but

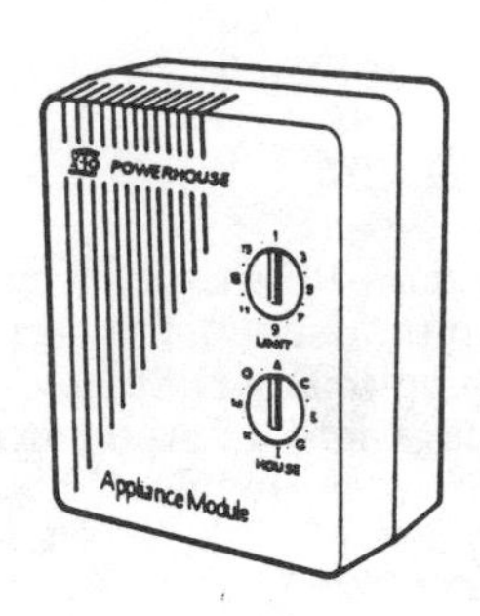

cannot dim lamps.

The Appliance Module does NOT respond to the "All Lights On" code but does respond to the All Units Off code.

The Appliance Module's "local control" feature lets you, at any time, turn on the appliance plugged into the Appliance Module simply by turning the appliance's power switch off then on again. You must always leave the power switch in the on position to control the Appliance Module remotely, but this handy local *on* feature means you don't have to always go to where the Controller is located to turn the lamp or appliance on. This feature does not work with "instant on" or remote controlled TVs.

APPLICATIONS

- The Appliance Module can be used to control many types of appliances such as a TV, stereo, radio, coffee pot, room air conditioner, toaster, toaster oven, humidifier, de-humidifier, lawn sprinkler, bug zapper, heated curlers etc.

- When used with X-10 Timers, the Appliance Module can turn on your stereo to wake you up to music, turn on the coffee pot so that you have fresh coffee when you wake up, and can automatically turn the coffee pot off after you have left for work. You can fall asleep at night to the TV which will automatically turn off at the time you set. Your TV and stereo can automatically turn on and off during the day to make your home sound "lived in". When used with the Thermostat Set-back Controller and the Telephone Responder, the Appliance Module lets you control your central heating and air conditioning in a second home from any phone, anywhere in the world!

- The Appliance Module can also be used to control a transformer or a relay which in turn can control something which requires a low voltage input or a contact closure such as a lawn sprnkler, an auto-dialer, drapery controls, window controls, automatic bed position controls, etc.

Wall Switch Module

Model No. WS467

The WS467 Wall Switch Module is rated for INCANDESCENT lights only, up to a max. rating of 500 watts. It can be turned on and off, and dimmed and brightened remotely from any X-10 Controller or Timer.

The WS467 responds to "All Lights On" from any Controller set to its Housecode. This allows you to turn **on** ALL lights connected to Wall Switch Modules at the touch of a button if you hear a strange noise at night, regardless of what Unit Code the Wall Switch Module is set to.

The WS467 also responds to "All Units Off" from any Controller set to its Housecode. This allows you to turn **off** ALL lights connected to Wall Switch Modules at the touch of a button when you go to bed, regardless of what Unit Code the Wall Switch Module is set to.

The WS467 has a push button for manual control which allows you to manually turn the light connected to it ON or OFF at the touch of a button. You can therefore always control the Module remotely whether or not you have turned it on or off locally.

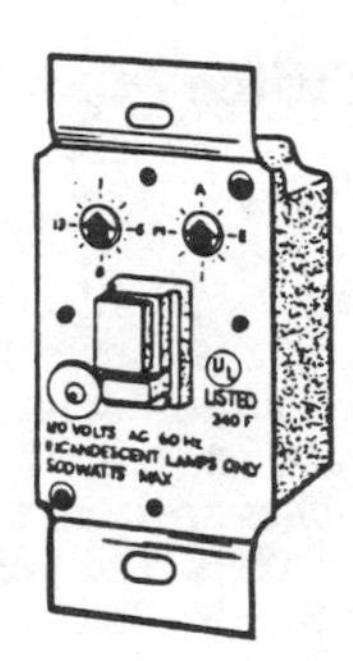

There is a slide switch located under the local control push button. This switch is a U.L. requirement and is used to turn the power off when changing a light bulb.

The WS467 has a "minimum" rating of 60W. If you try to control a fixture with a light bulb rated at less than 60 watts, the light bulb may glow dimly when the Wall Switch Module is turned off.

APPLICATIONS

- Use the WS467 to control Incandescent lights up to 500W max. (60W min.).
- Turn on lights from ANY X-10 Controller. The Mini Timer, Clock Radio Timer or the Home Automation Interface can turn on the WS467 at pre-set times to give your home a lived-in appearance.
- Use the Telephone Responder to turn on the Wall Switch Module from ANYWHERE IN THE WORLD! You can, for example, call your home and turn OFF lights which you forgot to turn off when you left the house. You could call home and turn ON the lights which you wish to be on when you arrive home, so that you don't have to enter a dark house.
- When used with the *POWERFLASH*™ Burglar Alarm Interface, the WS467 can FLASH any light connected to it when your alarm system trips.
- When used with X-10 Manual Controllers, the WS467 can provide "mood" lighting. You can DIM lights to suite your mood. You can even dim a group of lights simultaneously, a feature not possible with conventional dimmers. Simply press the buttons on the Maxi Controller corresponding to the codes of the Modules you wish to dim e.g. 1, 2, 4 and 12 and them press and hold the DIM button, Modules 1, 2, 4 and 12 will then ALL dim together.
- For the ultimate in mood lighting, the Home Automation Interface can even dim lights connected to Wall Switch Modules to a desired brightness level at a preprogrammed time.

3-Way Wall Switch Module

Model No. WS4777

The WS4777 Wall Switch Module is used to control a light which is presently controlled by two switches. The set includes a WS477 master switch and a CS277 companion switch. Both switches are used to replace the existing 3-way switches. The WS4777 is rated for INCANDESCENT lights only, up to a max. rating of 500 watts. It can be turned on and off, and dimmed and brightened remotely from any X-10 Controller or Timer.

The WS4777 responds to "All Lights On" from any Controller set to its Housecode. This allows you to turn **on** ALL lights connected to 3-Way Wall Switch Modules at the touch of a button if you hear a strange noise at night, regardless of what Unit Code the 3-Way Wall Switch Module is set to.

The WS4777 also responds to "All Units Off" from any Controller set to its Housecode. This allows you to turn **off** ALL lights connected to WS4777s at the touch of a button when you go to bed, regardless of what Unit Code the WS4777 is set to.

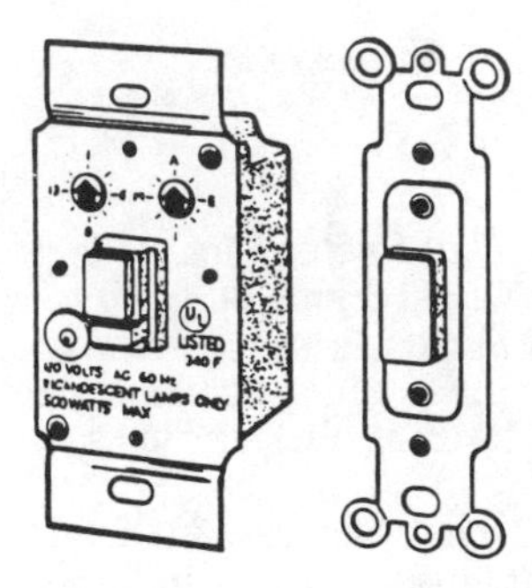

The WS477 and the CS277 have a push button for manual control which allows you to manually turn the light connected to them ON or OFF at a touch of the button on either switch. You can therefore always control the Modules remotely whether or not you have turned it on or off locally.

The slide switch under the local control push button on the WS477 is used to turn the power off when changing a light bulb.

The WS4777 has a "minimum" rating of 60W. If a light bulb rated at less than 60 watts is used, the light bulb may glow dimly when the Module is turned off.

APPLICATIONS

- Use the WS4777 to control Incandescent lights up to 500W max. (60W min.) which are presently controlled by two switches.
- Add another CS277 Companion Switch to control lights which are presently controlled by three switches.
- Turn on lights from ANY X-10 Controller. The Mini Timer, Clock Radio Timer or the Home Automation Interface can turn on the WS4777 at pre-set times to give your home a lived-in appearance.
- Use the Telephone Responder to turn on the WS4777 from ANYWHERE IN THE WORLD! You can, for example, call your home and turn OFF lights which you forgot to turn off when you left the house. You could call home and turn ON the lights which you wish to be on when you arrive home, so that you don't have to enter a dark house.
- When used with the *POWERFLASH*™ Burglar Alarm Interface, the WS4777 can FLASH any light connected to it when your alarm system trips.
- When used with X-10 Manual Controllers, the WS4777 can provide "mood" lighting. You can DIM lights to suite your mood. You can even dim a group of lights simultaneously, a feature not possible with conventional dimmers. Simply press the buttons on the Maxi Controller corresponding to the codes of the Modules you wish to dim e.g. 1, 2, 4 and 12 and them press and hold the DIM button, Modules 1, 2, 4 and 12 will then ALL dim together.

INSTALLING 3-WAY WALL SWITCH MODULES

- Disconnect the power at the main circuit breaker panel or fuse box.
- Remove the switch plates and unscrew the existing switches from their boxes.

Note that each existing switch has one screw terminal which is a different color from the other two terminals, this is the common terminal.

1. To replace one of the existing switches with the **Master** Switch - WS477 (the larger of the two switches):

- Select one of the existing switches (either one) and disconnect the wire on the existing switch from the terminal which is a different color from the other two terminals (the common terminal). Connect this wire to the BLUE wire on the **Master** Switch (using a wire nut).
- Remove the other two "traveller" wires from the existing switch and connect them in turn to the two remaining wires on the **Master** Switch.

Note: If one of the wires on the old switch is a RED wire, connect it to the RED wire on the **Master** Switch. In any case make a note of the color of the wire you connected to the RED wire on the **Master** Switch.

2. To replace the other existing switch with the **Companion** Switch - CS277 (the smaller of the two switches):

- Disconnect the wire on the existing switch from the terminal which is a different color from the other two terminals (the common terminal). Connect this wire to one of the two BLUE wires on the **Companion** Switch (using a wire nut).
- Remove the other two wires from the old switch and note the color of the wire you connected to the RED wire on the **Master** Switch. Connect this same wire to the RED wire on the **Companion** Switch.
- Connect the remaining wire from the old switch to the second BLUE wire on the **Companion** Switch.
- Install both switches into their wall boxes using the screws provided.
- Using a small screwdriver, set the House Code dial on the **Master** Switch to the same letter as the rest of your system.
- Set the Unit Code dial to the desired number (between 1 and 16).
- Move the power switch on the **Master** Switch to the on position (in the center).
- Re-fit the cover plates.
- Turn the power back on at the main circuit breaker panel or fuse box.

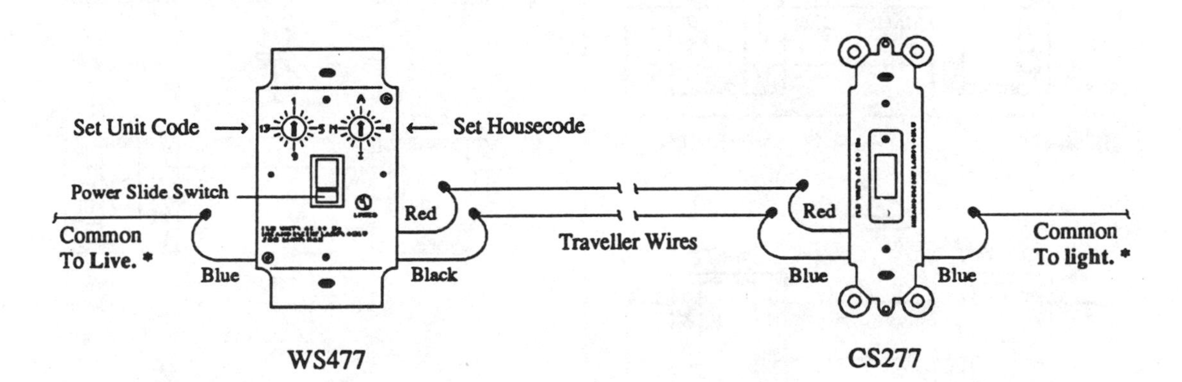

* Note: The WS477 and CS277 can also be installed the opposite way around to that shown above. i.e. with the common on the WS477 connected to the light and the common on the CS277 connected to Live.

OTHER TYPES OF 3-WAY AND 4-WAY INSTALLATIONS

The term 3-way (sometimes called 2-way) refers to a pair of wall switches connected so that both switches control the same light. For example, at the top and bottom of a staircase. In such a circuit, both switches have 3 connections. There are generally 2 "traveller wires" between the switches. The common terminal on one of the switches is connected to the live "hot" wire from the circuit breaker panel and the common terminal on the other switch goes to the light.

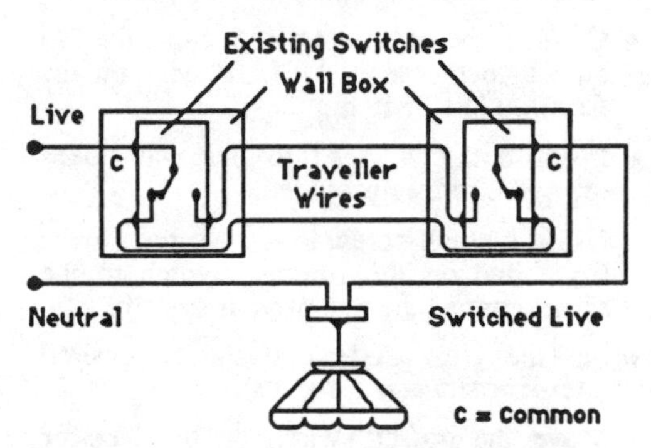

The diagrams below are X-10 equivalents of the circuit above. The Master and Companion switch can be installed at either end.

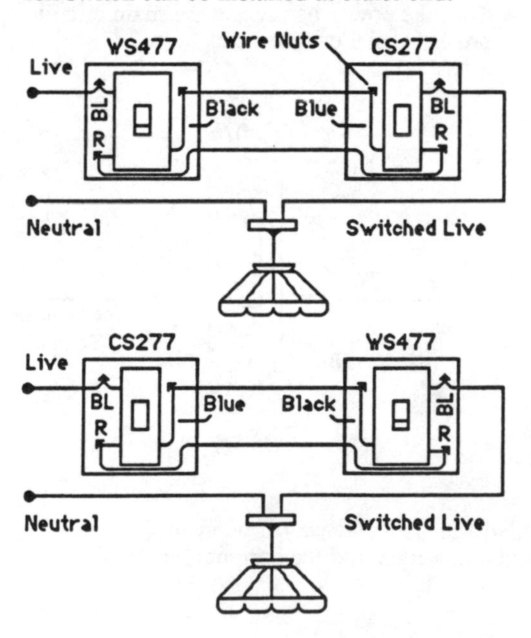

The preceding diagrams are "electrical" diagrams. In practice the actual physical connects can be made in 3 different ways, as shown below together with their X-10 equivalents.

1. Light fixture at end of run:

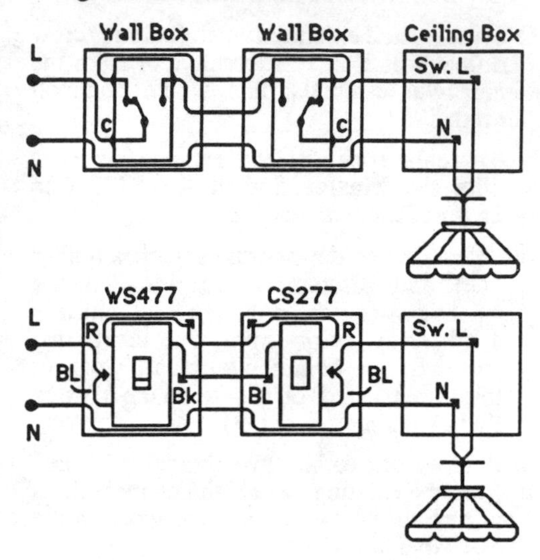

2. Light fixture in the middle of a run:

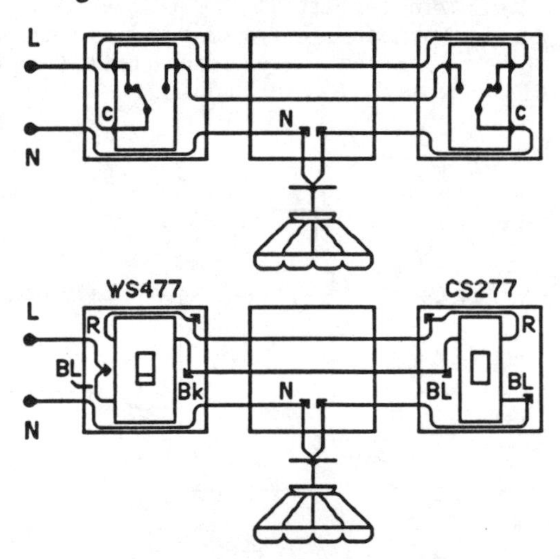

In any of these cases, the WS467 and the CS227 can be interchanged.

3. Light fixture at the start of a run:

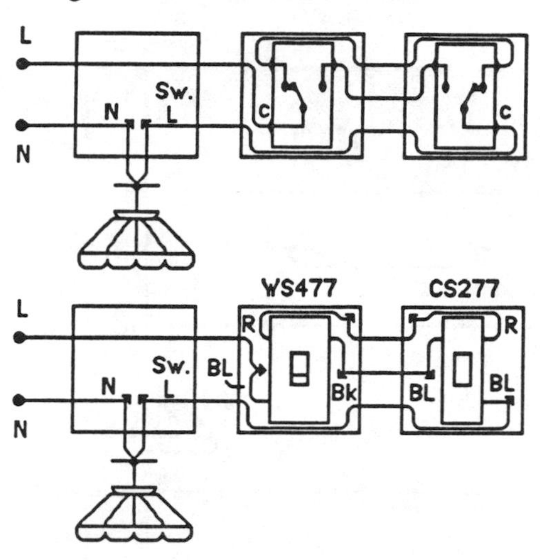

4-Way Wall Switch Circuit.

A 4-Way circuit is a circuit which has 3 switches controlling the same light. This circuit consists of two 3-way switches, one at each end of the run, and a 4-way switch in the middle. Operation of the 4 terminal, 4-Way switch reverses the connection of the two traveller wires, thus turning the light off if it was on or vice versa. See below.

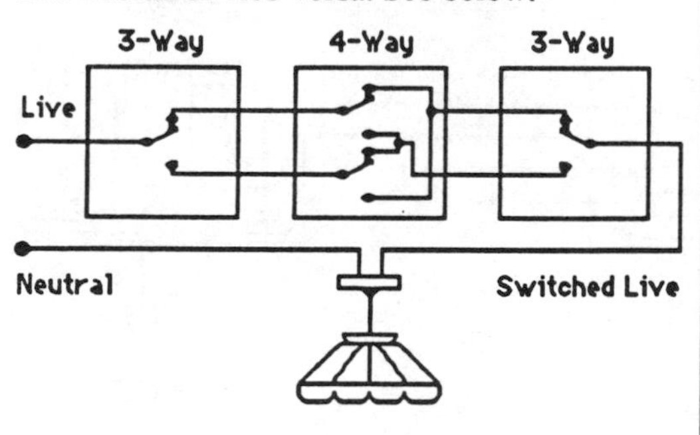

A typical implementation of this circuit is shown below.

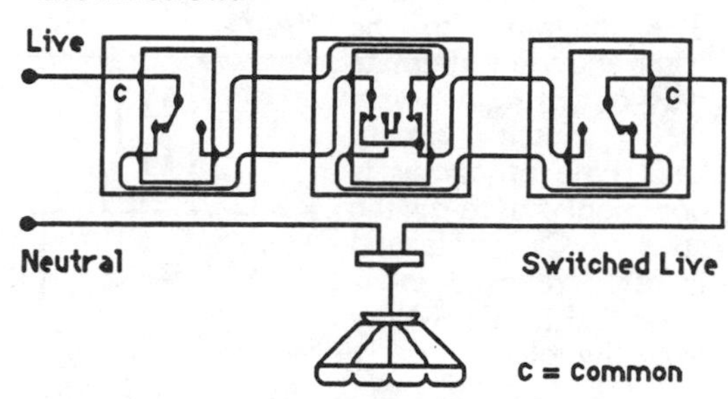

The 4-Way circuit above is replaced by X-10 switches by first replacing each 3-Way switch as you would for a regular 3-Way circuit. You then remove the 4-Way switch and connect the two traveller wires coming into the wall box to the two traveller wires coming out of the wall box, i.e. connect them straight through. Having done this, the circuit should operate as a regular 3-Way circuit from both of the 3-Way switches. You should check this before continuing.

You then replace the 4-Way switch with an additional companion switch (CS227) by connecting the two blue wires together and connecting both blue wires to one of the travellers and the red wire to the other traveller wire. See below.

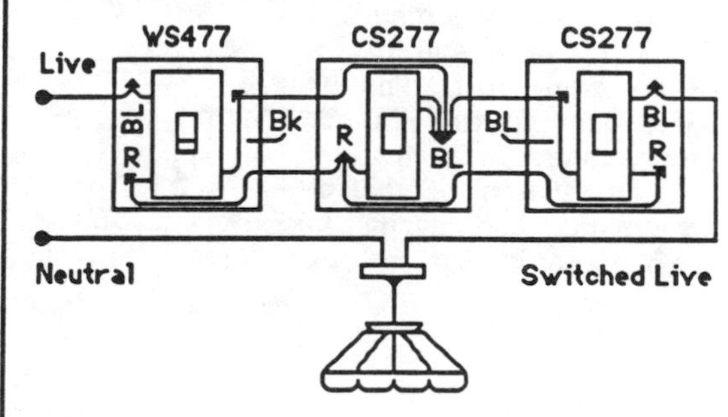

The California 3-Way Wall Switch

The diagrams to the right show an alternate way of wiring a 3-way wall switch circuit. This type of circuit is commonly used in the state of California and is often referred to as "The California 3-Way Wall Switch"

This circuit has the disadvantage of requiring 4 traveller wires but has the advantage that it provides live and neutral at both switch locations. Therefore, this type of circuit is commonly used where an AC outlet is required at the switch location. The circuit can be found with one or two light fixtures and with an outlet at either or both ends of the run.

The first and second diagrams show the schematic diagram and its X-10 equivalent. The third and fourth diagrams show a typical implementation of the schematic, and again, its X-10 equivalent.

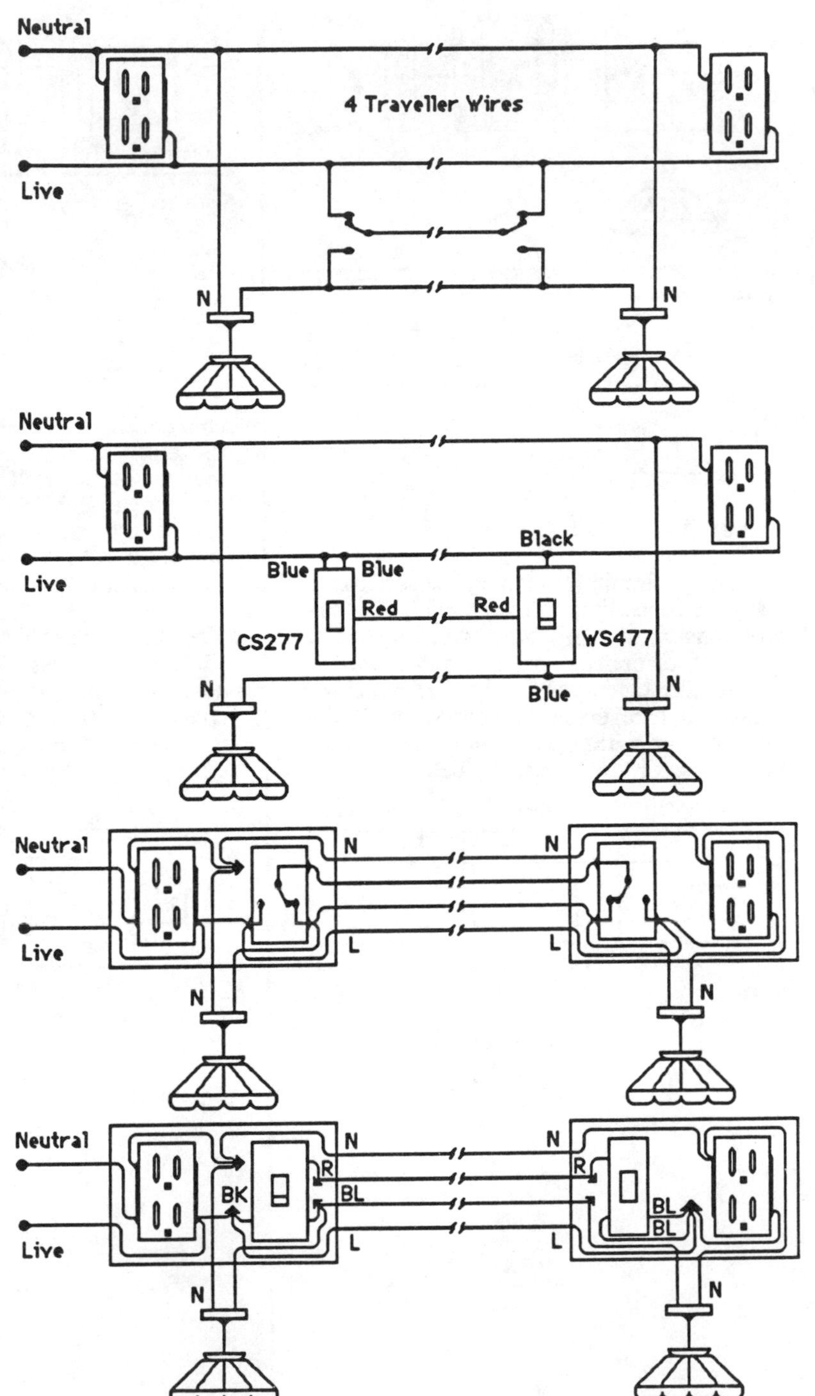

Split Receptacle Module

Model No. SR227

The SR227 replaces your existing wall receptacle. It has two outlets, the top outlet is controlled and works just like an Appliance Module. The bottom outlet is on all the time and works just like a regular outlet.

Both outlets are rated at 15A unconditionally. This is the equivalent of about 1800 watts so the SR227 will control ANYTHING that you would normally plug into a regular 15A wall receptacle.

Warning: extreme care should be taken when using the Split Receptacle Module to control portable heaters as a fire could result if the heater were turned on by remote control while clothing was draped nearby.

The Split Receptacle Module can be used for any kind of lamp including fluorescent lamps but cannot dim lamps.

The Split Receptacle Module does NOT respond to the "All Lights On" code but does respond to the All Units Off code.

The Split Receptacle Module's "local control" feature lets you, at any time, turn on the appliance plugged into the Split Receptacle

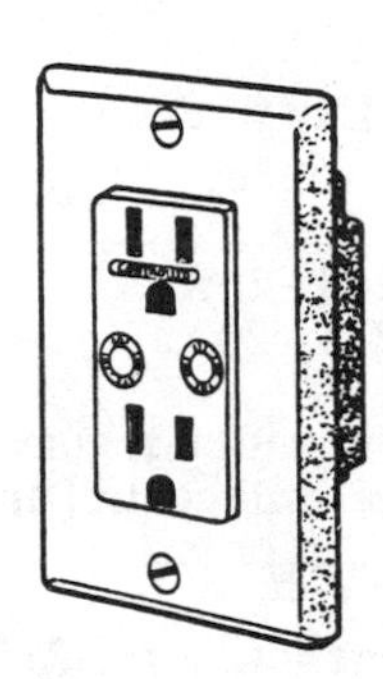

Module's top outlet simply by turning the appliance's power switch off then on again. You must always leave the power switch in the on position to control the Split Receptacle Module remotely, but this handy local *on* feature means you don't have to always go to where the Controller is located to turn the lamp or appliance on. This feature does not work with "instant on" and remote controlled TVs.

The SR227 Split Receptacle Module is packaged complete with a Decora™ style wall plate, mounting screws and wire nuts.

APPLICATIONS

- The Split Receptacle Module can be used to control many types of appliances such as a TV, stereo, radio, coffee pot, room air conditioner, toaster, toaster oven, humidifier, de-humidifier, lawn sprinkler, bug zapper, heated curlers etc.
- When used with X-10 Timers, the Split Receptacle Module can turn on your stereo to wake you up to music, turn on the coffee pot so that you have fresh coffee when you wake up and can automatically turn the coffee pot off after you have left for work. You can fall asleep at night to the TV which will automatically turn off at the time you set. Your TV and stereo can automatically turn on and off during the day to make your home sound "lived in". When used with the Thermostat Setback Controller and the Telephone Responder, the Split Receptacle Module lets you control your central heating and air conditioning in a second home from any phone, anywhere in the world!
- The Split Receptacle Module can also be used to control a transformer or a relay which in turn can control something which requires a low voltage input or a contact closure such as a sprinkler, an auto-dialer, drapery controls, window controls, automatic bed position controls etc.

220V Heavy Duty Appliance Module

Model No. HD243 - 15A
Model No. HD245 - 20A

The HD243 plugs into a regular 220V 15A or 20A outlet and will control any 220V appliance rated at 15A or less.

The HD245 plugs into a regular 220V 20A outlet and will control any 220V appliance rated at 20A or less.

Both Modules are designed to work on single split phase 110/220V or 120/240V systems. This is the kind of wiring system found in most houses. The HD243/245 will not work on 3 phase systems which are sometimes found in apartments.

This is because the signals transmitted by X-10 Controllers (onto 120V) are transmitted at the zero crossing point of the 120V line. All 120V Modules look for a signal at zero crossing. However, there is a 30 degree phase shift between zero crossing at 120V and zero crossing at 208V (phase to phase voltage in a 3 phase system is actually 208V not 220V). This 30 degree phase shift means that there is no signal present at the zero crossing point of the 208V phases where the 220V module is connected.

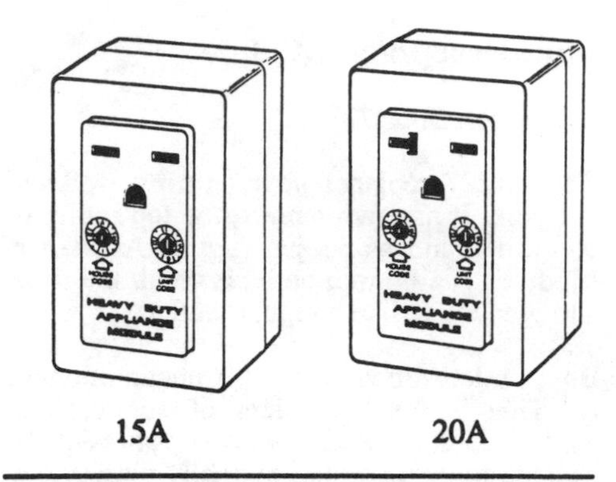

15A 20A

There is a product called a Repeater (model No. 6272) available to solve this problem when it is desired to use 220V Modules in industrial installations. The Repeater is used for large installations and is connected to all three phases in the breaker panel. It looks for a signal on any 120V phase and retransmits it at the correct zero crossing points onto all three phases.

For safety, the Heavy Duty Appliance Module does NOT respond to the "All Lights On" code but does respond to the All Units Off code.

The Heavy Duty Appliance Module does not have local control.

APPLICATIONS

- The 220V Heavy Duty Appliance Module can be used to control 220V appliances such as large window air conditioners or water heaters.
- Any X-10 Timer can turn off a water heater connected to a 220V Heavy Duty Appliance Module at times when hot water is not required. You can substantially reduce your energy bills by turning water heaters off periodically.
- When used with the Telephone Responder, the 220V Heavy Duty Appliance Module lets you control your large window air conditioner in a second home from any phone, anywhere in the world! Or, you can call home and turn on the air conditioner before you leave work so the house is nice and cool when you arrive home.
- Use the 220V Heavy Duty Appliance Module to turn on the heater in your hot tub so it's nice and warm by the time you get home.

Universal Module

Model No. UM506

The Universal Module contains an isolated contact relay and a piezo sounder (beeper). It can be set to operate as a beeper only, as an isolated contact relay only, or both.

It can be set for momentary or continuous operation for the sounder and/or isolated contact relay.

The isolated relay contacts are rated for 5A at 24VDC. Higher voltages can be switched (15A resistive at 120VAC, 500W for incandescent lamp load, 1/3 HP Inductive for Motors) but care must be taken to ensure that hazardous voltages are not present at the exposed screw terminals.

When set for continuous operation the relay contacts close and/or the module beeps continuously when an "on" command is received. When an "off" command is received, the relay contacts open and/or the beeps stop.

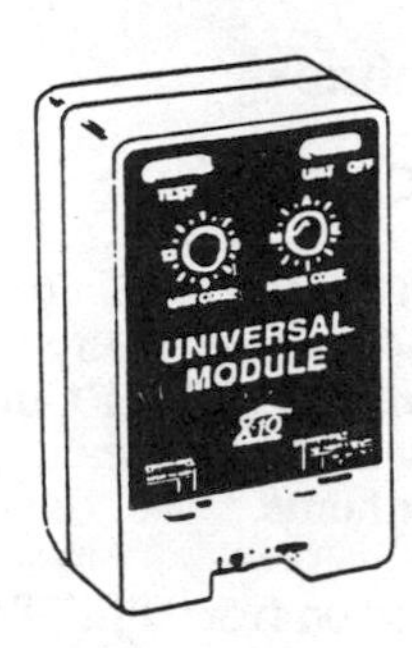

When set for momentary operation the relay contacts close and/or the beeper gives three or four beeps when an "on" command is received. The relay contacts then open automatically after about 1/2 second and/or the beeper stops automatically after three or four beeps. In this mode, off commands are ignored.

The module has a Test (On) button and a Unit Off button.

APPLICATIONS

- As a Sounder Module the UM506 can be set to give continuous beeps and could, for example, be used as a wake-up alarm when set to turn on from any X-10 Timer.
- It could also be set for momentary beeps and could be used as an annunciator - to call the children for dinner for example.
- It could be turned on from any X-10 compatible motion detector to announce that someone has approached your front door.
- When used as an isolated contact relay module, it can be set for a momentary or sustained contact closure. It could be used to operate drapery controls, window openers, garage door openers, gate operators, etc.
- It could also switch low voltages. You could use it to activate a 24V sprinkler system, a thermostat or anything which is operated by low voltage.
- Voltages higher than 24V can be switched but care must to taken as the terminals of the relay are exposed. If voltages higher than 30VAC are connected to the screw terminals on the module, the terminals must be labelled as hazardous, must be insulated from touch, and the Module must be installed in an inaccessible place or behind a locked cover.

Remote Chime

Model No. SC546

The *X-10 POWERHOUSE* Remote Chime (model SC546) has many applications. When used with the PR511 Outdoor Motion Detector, it will chime when someone approaches your home.

It can be turned on from a PF284 Powerflash Interface (which is triggered by a contact closure). Therefore if the contact closure is provided from a door-bell push button switch, the SC546 can be used as a remote door bell.

It can be programmed to chime from an MT522 Mini Timer or CP290 Computer Interface, to be used as a wake-up alarm.

It can be programmed from the CP290 to chime every hour, on the hour, to keep you aware of the time of day.

It can be used as a remote pager when turned on from any X-10 Controller (to signal to the kids that dinner is ready for example).

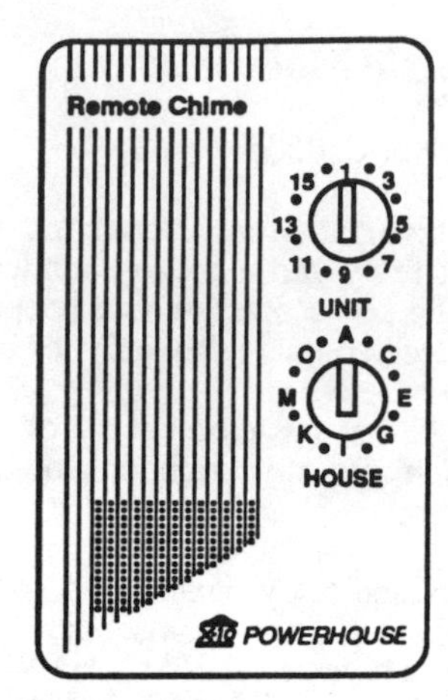

- Can be turned on from the PR511 Outdoor Motion Detector to announce that someone has approached your front door.
- Can be used as a wake-up alarm when set to turn on from any X-10 Timer.
- Can be used as an annunciator - to call the children for dinner for example.

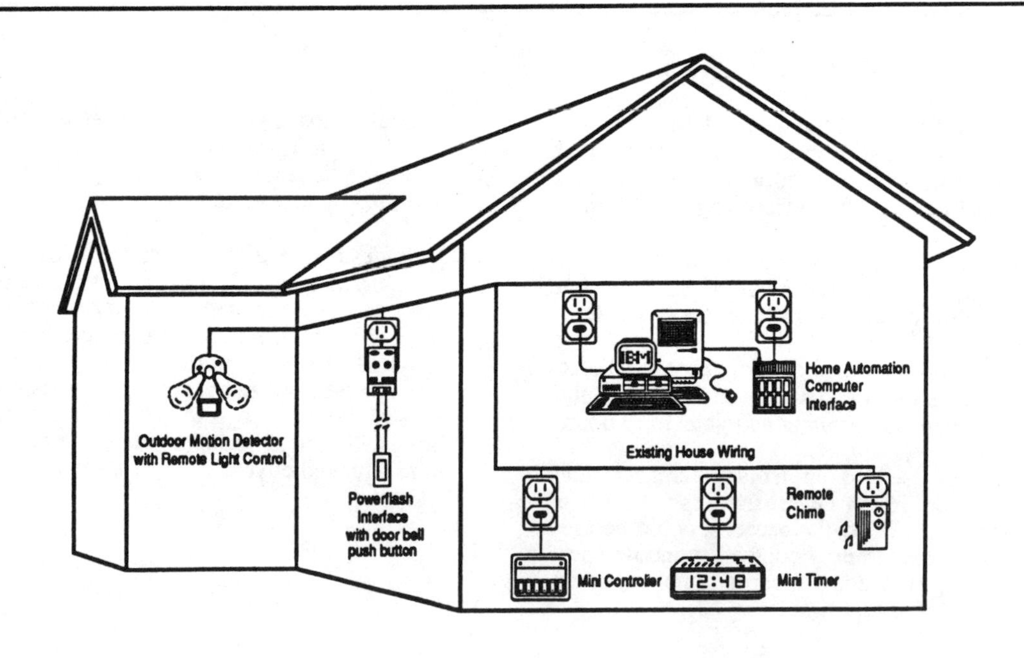

Appendix B

HM Sourcelist

AAMSCO Manufacturing, Inc.
P.O. Box 15119
Jersey City, NJ 07305
800/221-9092
(Lighting products)

Abloy Security
P.O. Box 35406
6200 Denton Dr.
Dallas, TX 75235
214/358-4762
(High-security cylinders and locks)

Ademco
165 Eileen Way
Syossett, NY 11791
800/645-7568 (ext. 1144)
516/921-6704 (ext. 1144)
(Alarm control panels and sensors; professional alarms)

Aiphone Corporation
1700 130th Ave. N.E.
P.O. Box 90075
206/455-0510
(Video intercoms)

Alarm Accessory Ltd.
P.O. Box 212
Claymont, DE 19703
(Alarm installation supplies)

Almont Lock Company, Inc.
113 School St.
Almont, MI 48003
313/798-8950
(Padlocks)

Alpine Electronics of America
Mobile Electronics Division
19145 Gramercy Pl.

Torrance, CA 90501
800/257-4631
213/326-8000
(Car alarms)

American Home Lighting Institute
435 N. Michigan Ave.
Chicago, IL 60611
312/644-0828
(Trade association)

Andersen Windows Inc.
P.O. Box 3900
Peoria, IL 61614
800/426-4261
(Doors; windows)

Arrow Fastener Company, Inc.
271 Mayhill St.
Saddle Brook, NJ 07662
201/843-6900
(Alarm installation supplies)

ASECO American Auto Security
236 E. Star of India Lane
Carson, CA 90746
800/421-1096
213/538-4670
(Car alarms)

Atrium Door & Windows Company
P.O. Box 226957
Dallas, TX 65222-6957
800/527-5249
(Doors; windows)

Benchmark, General Products
P.O. Box 7387
Fredericksburg, VA 22404
703/898-3800
(Doors)

Bennett Industries, Inc.
530 Palisade Ave.
Fort Lee, NJ 07024
201/947-5340
(Doors)

Bilco Company
P.O. Box 1203
New Haven, CT 06505
203/934-6363
(Doors)

BRK Electronics
780 McClure Rd.
Aurora, IL 60504-2495
708/851-7330
(Fire extinguishers; floodlights; smoke detectors; timers)

Brookfield Industries, Inc.
P.O. Box 548
Thomaston, CT 06787-0548
(Door hinges)

Buddy Products
1350 South Leavitt
Chicago, IL 60608
800/886-8688
(Wall safes)

Canadian Centre for Justice Statistics
Information and Client Services
19th Floor, R.H. Coats Building
Tunney's Pasture
Ottawa, Ontario
Canada K1A OT6
(Government agency)

Cannon Safe
9358 Stephens St.
Pico Rivera, CA 90660
800/242-1055
(Safes)

Caradco
201 Evans Rd.
Rantoul, IL 61866
217/893-4444
(Doors; windows)

Carbrella Motoring Accessories
7000 E. Quincy Ave., #116F
Denver, CO 80237
303/290-0577
(Steering-wheel locks)

CHB Industries
92 Lakewood Ave.
Ronkonkoma, NY 11779
516/981-2746
(Security film)

Clifford Electronics, Inc.
20750 Lassen St.
Chatsworth, CA 91311
800/824-3208
818/709-7551
(Car alarms; stolen-vehicle retrieval systems)

CodeAlarm
32380 Edward Ave.
Madison Heights, MI 48071
800/468-3723
313/583-7460
(Car alarms; stolen-vehicle retrieval systems)

Cole Sewell Corporation
2288 West University Ave.
St. Paul, MN 55114
800/328-6596
(Doors; windows)

Compu-Gard
36 Maple Ave.
Seekonk, MA 02771
800/333-6810
508/761-4520
(Computer security hardware)

Crestline
One Wausau Center
P.O. Box 8007
Wausau, WI 54402-8007
800/552-4111
(Doors)

Crime Prevention Coalition
P.O. Box 6600
Rockville, MD 20850-0635
(Consumer information)

Crimestopper Security Products, Inc.
1770 S. Tapo St.
Simi Valley, CA 93063
800/998-6880
805/526-9400
(Car alarms)

David Levy Company, Inc.
Box 6033
Artesia, CA 90702-6033
310/404-9998
(Car alarms)

DesignTech International, Inc.
7401 Fullerton Rd.
Springfield, VA 22153
703/866-2000
(Car alarms; home automation systems)

Dicon Systems Ltd.
719 Clayson Rd.
Toronto (Weston), Ontario
Canada M9M 2H4
416/745-6044
(Home alarms)

Dimango Products Corporation
7258 Kensington Rd.
Brighton, MI 48116
313/486-0770
(Home alarms; video intercoms)

Directed Electronics, Inc.
2560 Progress St.
Vista, CA 92083-8422
800/274-0200
(Car alarms)

Diversified Manufacturing & Marketing
Company, Inc.
1207 Grant Rd.
Graham, NC 27253
800/672-5885
919/227-7012
(Alarm installation supplies)

DOM Security Locks
100 Central Ave.
Brockville, Ontario K6V 5W6
Canada
800/363-4803
613/342-6641
(High-security cylinders)

Don-Jo Manufacturing
P.O. Box 929
Sterling, MA 01564
800/628-8389
508/422-3377
(Door hardware; strike boxes)

Door Systems Inc.
333 Byberry Rd.
Hatboro, PA 19040
215/672-8087
(Pushbutton door locks)

Eastman Wire & Cable Company
1085 Busch Memorial Hwy.
Pennsauken, NJ 08110
800/257-7940
609/488-8800
(Alarm installation supplies)

Excalibur of America
P.O. Box 508
Douglasville, GA 30133
404/942-9876
(Car alarms)

Falcon Eye, Inc.
3130 Marquita Dr.
Fort Worth, TX 76116
817/244-2860
(Outdoor lighting)

Federal Crime Insurance Program
P.O. Box 6301
Rockville, MD 20849-6301
800/638-8780
(Federal agency)

Fichet-Brauner, USA
1255-A Oakbrook Dr.
Norcross, GA 30093
800/582-0696
404/448-5593
(High-security cylinders and locks)

Fort Knox
1051 N. Industrial Park Rd.
Orem, UT 84057
800/821-5216
801/224-7233
(Fire safes)

Gardall Safe Corporation
P.O. Box 30
Eastwood Station
Syracuse, NY 13206
800/722-7233
315/432-9115 (New York)
(Fire and burglary safes)

GE Lighting
Nela Park
Cleveland, OH 44112
800/626-2000
(Lighting products)

George Kovacs Lighting Inc.
67-25 Otto Rd.
Glendale, NY 11385
718/392-8190
(Lighting products)

Generation Two
10777 Barkley, Suite 105
Overland Park, KS 66211
913/642-1518
(Smoke detectors)

Group Three Technologies, Inc.
2125-B Madera Rd.
Simi Valley, CA 93065
805/582-4410
(Home automation systems)

GTE Electrical Products
100 Endicott St.
Danvers, MA 01923
(Lighting products)

Harrison Electronics Systems, Inc.
Box 1758
Wilkes-Barre, PA 18705
800/422-5050
717/825-0540
(Car alarms)

Heath Zenith Reflex Brand Group
455 Riverview Dr.
Benton Harbor, MI 49022
616/925-5181
(Do-it-yourself home alarms)

Hidden Assets
P.O. Box 20056
Cherokee Station
New York, NY 10028-0050
("Hidden-pocket" clothing)

Home Automation, Inc.
P.O. Box 9310
Metairie, LA 70055-9310
504/833-7256
(Home automation systems)

Home Mechanix Magazine
Two Park Ave.
New York, NY 10016
212/779-5000
(Home/Car improvement magazine)

Honeywell, Inc.
Home Building Controls
1985 Douglas Drive N.
Golden Valley, MN 55422
612/870-2926
(Door alarms; home automation systems)

Idaho Wood
P.O. Box 488
Sandpoint, ID 83864
800/635-1100
(Outdoor lighting)

InteLock Corporation
7026 Koll Center Parkway, Suite #225
Pleasanton, CA 94566
510/462-2114
(Electronic door locks)

Intermatic Inc.
Intermatic Plaza
Spring Grove, IL 60081-9698
815/675-2321
(Timers)

International Association of Home Safety and Security Professionals
P.O. Box 2044
Erie, PA 16512-2044
(Consumer information; referral service; trade association)

International Association of Lighting
Designers
18 E. 16th St.
New York, NY 10013
212/206-1281
(Trade association)

Jameson Home Products
2820 Thatcher Rd.
Downers Grove, IL 60515
800/445-8299
(Home security products)

Jeld-Wen, Inc.
Commerce Dr.
Mount Vernon, OH 43050
800/535-3963
(Doors; windows)

Jessup Door Company
300 E. Rail Rd.
Dowagiac, MI 49047
800/826-2367
(Doors)

Johnson Products, Inc.
P.O. Box 1126
Elkhart, IN 46515
219/293-5664
(Doors)

Karsulyn Corporation
542 Industrial Dr.
Lewisberry, PA 17339
800/932-EXIT
717/938-0256
(Escape equipment)

Kenwood
2201 E. Dominguez St.
Long Beach, CA 90810
310/639-9000
(Car alarms; stolen-vehicle retrieval systems)

Kryptonite Corporation
320 Turnpike St.
Canton, MA 02021
617/828-6655
(Steering-wheel locks)

Kwikset Corporation
516 E. Santa Ana St.
Anaheim, CA 92805
714/535-8111
(Mechanical door locks; padlocks)

Latch-Gard
P.O. Box 425
Wakarusa, IN 46573
219/862-2373
(Door hardware)

Leslie-Locke, Inc.
P.O. Box 723727
Atlanta, GA 30339
404/953-6366
(Bars; doors; grilles; windows)

Lightoiler, Inc.
100 Lighting Way
Secaucus, NJ 07096
201/864-3000
(Indoor and outdoor lighting)

Linear
2055 Corte Del Nogal
Carlsbad, CA 92009
619/438-7000
(Home alarms and sensors; video intercoms)

LoJack Corporation
333 Elm St.
Dedham, MA 02026
617/444-4900
(Stolen-vehicle retrieval systems)

M.A.G. Engineering & Manufacturing, Inc.
15261 Transistor Lane
Huntington Beach, CA 92649
714/891-5100
(Door and window hardware; strike boxes)

Malta Company, The
P.O. Box 397
Malta, OH 43758
614/962-3131
(Doors; windows)

Markar Products, Inc.
12715 Lewis Rd.
Akron, NY 14001
800/866-1688
716/542-3001
(Door hinges)

Marvin Windows & Doors
P.O. Box 100
Warroad, MN 56763
800/346-5044
(Doors; windows)

Master Lock Company
2600 N. 32nd St.
Milwaukee, WI 53210
414/444-2800
(Mechanical door locks; padlocks)

Medeco High Security Locks
P.O. Box 3075
Salem, VA 24153
703/380-5000
(High-security cylinders and locks)

Meister Atlanta Corporation
3673 Clairmont Rd.
Atlanta, GA 30341
404/451-9700
(Strike boxes)

Monsanto Chemical Company
800 N. Lindbergh Blvd.
St. Louis, MO 63167
314/694-1000
(Security film)

Morgan Manufacturing
P.O. Box 2446
Oshkosh, WI 54903
800/766-1992
(Doors)

Moultrie Manufacturing Company
P.O. Drawer 1179
Moultrie, GA 31776-1179
800/841-8674
912/985-1312
(Fences and gates)

MSI Mace
P.O. Box 496
Hoosick Falls, NY 12090
(Door braces; mace; security devices)

Mul-T-Lock Corporation
54-45 44th St.
Maspeth, NY 11378
800/622-LOCK
718/392-9000
(High-security cylinders and door locks)

Nana Windows & Doors
707 Redwood Highway
Mill Valley, CA 94941
800/873-5673
415/383-3148
(Doors; windows)

Napco Security Systems, Inc.
333 Bayview Ave.
Amityville, NY 11701
800/645-9445
516/842-9400
(Alarm control boxes; sensors; professional home alarm systems)

National Association of Private Security Vaults
3562 N. Ocean Blvd.
Ft. Lauderdale, FL 33308
305/565-7466
(Trade association)

National Fire Information Council
P.O. Box 23221
Lansing, MI 48909
517/655-5355
(Consumer information)

National Fire Sprinkler Association, Inc.
P.O. Box 1000
Patterson, NY 12563
914/878-4200
(Trade association; consumer information)

New England Lock & Hardware Company, The
P.O. Box 544
Norwalk, CT 06854
203/866-9283
(Mechanical cylinders and door locks)

Niland Company
7241 Styles
El Paso, TX 79915
915/779-1405
(Lighting products)

Norco Windows, Inc.
P.O. Box 140
Hawkins, WI 54530
715/585-6311
(Windows)

Norden Lock Company, Inc.
36J Carlough Rd.
Bohemia, NY 11716
516/563-0900
(High-security mechanical door locks)

Novi International
9424 Abraham Way
Santee, CA 92071
619/258-1500
(Video intercoms)

NuTone
Madison and Red Banks Roads
Cincinnati, OH 45227-1599
800/543-8687
800/582-2030 (Ohio)
(Video intercoms)

Peachtree Windows & Doors
4350 Peachtree Industrial Blvd.
Norcross, GA 30071
404/497-2000
(Doors; windows)

Pease Industries
7100 Dixie Highway
Fairfield, OH 45014
513/870-3619
(Doors)

Pella Rollscreen Company
102 Main St.
Pella, IA 50219
800/524-3700
(Doors; windows)

Perma-Door Manufacturing
9017 Blue Ash Rd.
Cincinnati, OH 45242
513/745-6400
(Doors)

Philips Lighting Company
P.O. Box 6800
Somerset, NJ 08875
201/563-3000
(Lighting products)

Pinecrest, Inc.
2118 Blaisdell Ave.
Minneapolis, MN 55404
800/443-5357
(Door and window hardware; doors; windows)

Pittsburgh Corning Corporation
800 Presque Isle Dr.
Pittsburgh, PA 15239
800/642-2120
(Glass block)

Preso-Matic Lock Company
3048 Industrial 33rd St.
Fort Pierce, FL 34946-8694
407/465-7400
(Pushbutton door locks)

Profile Consumer Electronics
17211 S. Valley View Ave.
Cerritos, CA 90701
310/802-9007
(Video intercoms)

Progress Lighting
Erie Avenue G St.
Philadelphia, PA 19134
215/289-1200
(Indoor and outdoor lighting)

Questech International Inc.
4951 B.E. Adamo Dr., Suite 238
Tampa, FL 33605
813/247-4900
(Security telephones)

Quorum International, Ltd.
8777 East Via De Ventura, Suite 188
Scottsdale, AZ 85258
602/951-6990
(Personal alarms)

Rejuvenation Lamp & Fixture Company
1100 S.E. Grand Ave.
Portland, OR 97214
503/231-1900
(Indoor and outdoor lighting)

Rudolph-Desco Company, Inc.
580 Sylvan Ave.
Englewood Cliffs, NJ 07632-3105
(Door viewers)

Schlage Lock Company
2401 Bayshore Blvd.
San Francisco, CA 94134
415/467-1100
(High-security cylinders and door locks; mechanical door locks)

Seco-Larm USA Inc.
17811 Sky Park Circle, Suite D&E
Irvine, CA 92714
800/662-0800
714/261-2999
(Car alarms)

Secure-it, Inc.
18 Maple Court East
Longmeadow, MA 01028
800/451-7592
(Computer security hardware)

Sentry Group
900 Linden Ave.
Rochester, NY 14625
716/381-4900
(Fire safes; insulated file cabinets; media chests)

Simplex Access Controls
P.O. Box 4114
Winston-Salem, NC 27115-4114
919/725-1331
(Pushbutton door locks)

Simpson Door Company
P.O. Box 210
McCleary, WA 98557
206/495-3291
(Doors)

Sony Security Systems
15 Essex Rd., 3rd Fl.
Paramus, NJ 07652
201/368-5018
(CCTV cameras, monitors, and peripherals)

Stanley Door Systems
Div. of The Stanley Works
1225 E. Maple Rd.
Troy, MI 48083
(Doors)

Stanley Hardware
Div. of The Stanley Works
195 Lake St.
New Britain, CT 06050
800/622-4393
203/225-5111
(Door and window hardware)

Stor-A-Dor
P.O. Box 1661
Darien, CT 06820
203/655-6786
(Doors)

Taylor Brothers
P.O. Box 11198
Lynchburg, VA 24506-1198
804/237-8100
(Doors)

Techne Electronics Ltd.
916 Commercial St.
Palo Alto, CA 94303
800/227-8875
415/856-8646
(Car alarms)

Therma-Tru Corporation
1684 Woodlands Dr.
Mauwee, OH 43537
419/891-7400
(Doors)

Thomas Industries, Inc.
Residential Lighting Division
950 Breckenridge Lane
Louisville, KY 40207
502/886-3311
(Indoor and outdoor lighting)

Topper Hardware, Inc.
2999A Alhambra Dr.
Cameron Park, CA 95682
916/677-3001
(Door-lock parts and supplies)

Transcience
633 Hope St.
Stamford, CT 06907
203/327-7810
(Home alarms)

Twenty First Century
3249 W. Story Rd.
Irving, TX 75038
214/252-6201
(Fire extinguishers)

Ultrak
2400 Industrial Lane, Suite 2000A
Broomfield, CO 80020
800/846-8884
303/466-7333
(CCTVs; intercoms)

United States Department of Justice
Bureau of Justice Statistics
Washington, DC 20531
(Federal agency)

United States Fire Administration
Office of Fire Prevention and Arson Control
16825 South Seton Ave.
Emmitsburg, MD 21727
(Federal agency)

United States Postal Service
Postal Inspection Service
Washington, DC 20260-2186
(Mail fraud information)

Unity Systems Inc.
2606 Spring St.
Redwood City, CA 94063
415/369-3233
(Home automation systems)

Vantage Technologies
425 Pleasant St.
Fall River, MA 02721
508/678-2110
(Do-it-yourself home alarms)

Vehicle Security Electronics, Inc.
21540-F Prairie St.
Chatsworth, CA 91311
800/932-9999
818/700-7900
(Car alarms)

Velux-America, Inc.
P.O. Box 5001
Greenwood, SC 29648
800/283-2831
(Windows)

Vicon Industries Inc.
525 Broad Hollow Rd.
Melville, NY 11747
800/645-9116
516/293-2200
(CCTV cameras, monitors, and peripherals)

Vigilante Burglar Bars, Inc.
35 Country Ridge Dr. N.
Portchester, NY 10573
212/328-3700
(Windows bars; grilles; screens)

VSI Donner
12930 Bradley Ave.
Sylmar, CA 91342
800/882-0116
818/367-2131
(Door and window hardware; padlocks)

Watchguard Inc.
1001 Club Lakes Parkway
Lawrenceville, GA 30244
800/428-4889
404/279-1021
(Door and window hardware and locks)

Weather Shield Manufacturing
One Weather Shield Plaza
Medford, WI 54451
800/477-6808
(Doors; windows)

Wenco Windows
Box 1248
Mount Vernon, OH 43050
800/535-3936
(Windows)

Winner International
32 West State St.
Sharon, PA 16146
800/527-3345
800/344-2671
(Steering-wheel locks)

X-10 (USA) Inc.
91 Ruckman Rd.
Box 420
Closter, NJ 07624-0420
800/526-0027
201/784-9700
(Alarm sensors; do-it-yourself home alarms; home automation components and sensors; timers)

Yale Locks and Hardware
P.O. Box 25288
Charlotte, NC 28229-8010
800/438-1951
704/283-2101
(Mechanical cylinders and door locks)

Glossary

Activate To trigger an armed alarm system so that it detects sound, motion, smoke, heat, or other alarm conditions.

Active arming A type of car alarm that requires an action other than turning off the ignition or closing the doors to be armed.

A-lamp A technical term for the standard incandescent light bulb used in most homes.

Annunciator A signaling device—such as a bell, siren, or light—used in an alarm system to show that the system has been activated.

Antiscanning circuitry Receiver circuitry in a car alarm that detects when a scanner is being used in an attempt to disarm the system.

Arm To turn on a security system, enabling it to detect an alarm condition.

Audio discriminator A detection device that recognizes certain sounds, such as that of breaking glass.

Backset The horizontal distance from the lock edge of a door to the center of an installed cylinder or installed door knob.

Backup battery A standby battery available to be used as the primary power supply if necessary. A backup battery is usually a part of an alarm system.

Beveled edge A door edge that forms an angle of less than 90 degrees with the wide face of the door. One side of the beveled edge is the high side; the other is the low side.

Bolt The moving metal part of a lock; it enters a strike plate or strike box when in the locked position.

Brace lock A lock that uses a bar, angled between the door and floor, to wedge the door closed. Also called "buttress lock."

Burning bar A bar that is used to burn through safes. Also called "thermal lauce."

Central station A company that monitors electronic alarm systems. Also called "central monitoring station."

Certified master locksmith (CML) A certification granted by the Associated Locksmiths of America.

Certified professional locksmith (CPL) A certification granted by the Associated Locksmiths of America.

Certified protection officer (CPO) A certification granted by the International Association of Professional Security Consultants.

Certified protection professional (CPP) A certification granted by the American Society for Industrial Security.

Closed circuit television system (CCTV) A television system that limits reception of images to directly connected monitors.

Control panel The "brain" of an electronic alarm system.

Cylinder The cylindrical part of a lock, where the key goes into.

Cylinder guard A circular metal piece that fits around the cylinder of a lock to make the cylinder less vulnerable to wrenching and other types of attack. Also called "cylinder collar."

Deadbolt lock A type of lock with a bolt that locks into place once it's fully extended and can't be retracted by pushing against it. The most popular style of deadbolt is installed on a door by boring two connecting holes—one through the face of the door and one through the edge of the door. Also called "deadbolt" or "tubular deadbolt."

Detection device A device used in an alarm system to monitor and detect changes in a condition. Also called "sensor."

Digital communicator A device used in an electronic alarm system to dial one or more telephone numbers in order to transmit digitally coded information about the system's status. Also called "digital dialer."

Disarm To turn an alarm system (or specific zones) off so that the system (or the specific zones) won't respond to an alarm condition.

Double-cylinder lock A lock that has two cylinders.

Escutcheon plate Protective ornamental hardware used on a door.

Hardwired alarm system An alarm system that requires wires to be connected between its sensors and control panel. Also called "hardwire alarm system."

Header Horizontal member of door frame at the top of the opening.

High-Intensity Discharge (HID) Lamp A lamp (for light bulb) that uses gases within the bulb to produce light. All HID lamps require special fixtures and ballasts.

Hinge jamb The vertical member of a door frame, used for installing hinges.

Incandescent lamp A standard household light bulb; contains a metal filament and produces a yellow-white light.

Jamb A vertical member of a door frame.

Jimmyproof deadlock A surface-mounted lock with cylindrical bolts that move vertically to lock into the eye-loops of a compatible surface-mounted strike. Also called "jimmy-resistant deadbolt" or "surface-mounted dropbolt."

Key blank An uncut key that locksmiths use for duplicating keys. Also called "blank."

Key-in-knob lock A lock shaped like one or two door knobs with a keyway in one or both knobs. Also called "key-in-lock," "cylindrical lock," or "Lock-in-knob."

Lamp Lighting designers technical term for what most laypersons call a light bulb. A tube that uses electricity to produce light.

Low-voltage Lighting Lighting that operates on 12-volt or 24-volt current instead of 120-volt current; it may use a plug-in transformer to reduce household current to 12 volts or 24 volts.

Lumen A unit of measurement of illumination at a light source. One lumen equals the amount of light emitted by a standard candle.

Lux A metric unit of luminance; one lux equals one lumen per square meter.

PAR (parabolic aluminized reflector) lamp A type of incandescent light bulb designed for outdoor use.

Rail A horizontal piece of a window sash that connects with two stiles.

R-lamp Reflector lamp. A type of incandescent light bulb designed for indoor use.

Registered locksmith (RL) A certification granted by the Associated Locksmiths of America.

Residential Protection Specialist (RPS) A certification granted by the International Association of Home Safety and Security Professionals.

Shim A piece of wood or metal used to level or square a door or window.

Sill The horizontal part of a door frame that runs along the floor.

Spring-loaded bolt A beveled bolt that can be pushed back by applying pressure to it.

Stile A vertical piece of the framework of a wood door, or a vertical piece that connects with two rails of a window sash.

Strike box A strike plate with a metal or plastic box to surround the lock bolt.

Strike jamb The vertical member of a door frame used to install a strike plate or strike box. Also called "lock jamb."

Strike plate A metal plate with one or more bolt-hole openings; installed on a door jamb to accept the bolt or bolts of a lock. Also called "strike."

Stud A vertical wall member, usually 2" × 4" lumber, to which drywall or other wall covering is attached.

Threshold A raised member extending between the jambs of a door frame along the floor.

Transformer A device that transforms current of one voltage into current of another voltage. Transformers are used in low-voltage lighting and security systems to transform household current into 12 volts or 24 volts.

Wireless alarm system An alarm system that doesn't require wire to be connected between its sensors and control panel.

Zone A specific area (or group of areas) of protection in an alarm system that can be armed and disarmed independent of other areas of protection in the system.

Index

Abloy Disklock, 27
Abloy Security, Inc., 28–29, 31
Accidents, preventing, 86
Active alarm, 125
Aiphone Corporation, 98–99
Alarm Accessory Ltd., 74
Alarm system installers, 154
Alarms:
  audio, 65–66
  do-it-yourself, 68–70
  dual techs, 67–68
  foil, 64–65
  hardwire, 69
  infrared, 66–67
  magnetic, 65
  microwave, 66
  quads, 67
  self-contained, 69–70
  ultrasonic, 66
  wireless, 69, 71
ALOA, 154
American National Standards Institute (ANSI), 26
American Society for Industrial Security (ASIA), 156
Anti-theft devices, 124–125
Antilift plate, 10
Audio discriminators, 65

Bit-key, 26–27
BRK Electronics, 112
Burglar alarms, 64–81
Burglars:
  types of, 4–5

Cameras, 96
Car alarms, 125–129
Carbrella Motoring Accessories, 125
Carjackers, 130–132
Car protection, 131
Certified Protection Professional (CPP)
    certification, 156
CHB Industries, 22–23
Clean-up tips, 119
Closed circuit television, 96–104
  installing, 97–98
Combating carjackers, 130–132
Cylinder lock, 45–46

David Levy Company, Inc., 126
Deadbolt, 27–30
  installation, 9
Detection devices, 64
Do-it-yourself system, 68
Door and Hardware Institute, 154
Door brace, 140
Door frame, 9
Door Spy, Inc., 12
Doors, 7–17
  barriers, 10–12
  brace, 140
  breaking in, 8

Doors (*Continued*)
chains, 10
choosing, 15, 17
frame, 9
garage, 14–15
reinforcers, 10–11
replacing, 16
sliding glass, 10, 14
strength of, 17
wide-angle viewer, 10–12
Dual techs, 67

Electricity safety, 81
Electronic locks, 47–51
Elert, 139
Escape ladders, 112
Expert advice:
alarms, 79–81
closed circuit television, 106–107
lighting, 92–94

Falcon Eye, Inc., 87–90
Federal Crime Insurance Program, 146–149
Federal Insurance Administration, 146
File cabinet, 60
Fire:
causes, 110–111
clean-up, 119–120
detectors, 111
escapes, 113–118
extinguishers, 111–113
surviving, 118–119
threat, 5
Fire extinguishers, 111
Fire safe, 54
Floor safes, 55
Fluorescent lighting, 85
Foil, 64
Frank, Harold, 106

Garage doors, 14–15
reinforcing, 14
Gardall Safe Corporation, 56–57
Glass blocks, 20–21
installing, 21–22
Grenald Associates, 92
Guardian, The, 113, 116
Guardplate covers, 32

H. Frank & Associates, 106
Halogen, 85
Hardwired alarm system, 69
Hardwiring, 72
Heath Zenith Reflex Brand Group, 69, 88
Hidden Assets, 138
High-intensity discharge (HID), 85
High-security cylinders, 30
High-security strike box, 8
Hinge enforcers, 8
Hinges, removing, 8
Home automation controllers, 77
Home automation systems, 72

In-floor safes, 56
Infrared detectors, 66
Insurance, 144–149
disaster, 145
discounts, 145–140
how much, 144–145
personal, 145
understanding, 144
Insurance adjusters, 121
InteLock Corporation, 47, 51
Intermatic, Inc., 86
International Association of Home Safety and Security Professionals, Inc. (IAHSSP), 152, 156
International Association of Lighting Designers, 92
International Association of Professional Security Consultants (IAPSC), 156

Japanese Industrial Standards (JIS), 54
Jimmying, 28
Jimmyproof deadlock, 30–41

Kalamen doors, 15
Karsulyn Corporation, 114–115
Key-in-knob, 26–27
Key-in-lever, 26–27
Keys:
copies, 45
minimizing, 46–47
rekeying, 46–47
Kryptonite Corp., 124
Kwikset Corporation, 26, 45

Laminated glass, 20
Leslie-Locke, Inc., 22
Lexan, 20
Lexigard, 20
Light controllers, 85
Light sources, 85

Lighting, 84–94
  controllers, 85–86
  landscape, 84–85
  sources, 85
Lockpicking, 28
Locks:
  choosing, 27–28
  cylinder, 45–46
  deadbolt, 10
  electronic, 47–49
  grades, 26
  installing, 30–52
  jimmyproof, 30–32
  police, 42–44
  pushbutton, 47
  types, 26–52
Lockpicking, 28
Locksmiths, 153–154
LoJack Corporation, 130

M.A.G. Engineering & Manufacturing Co., 9–11, 14
Magnetic switches, 65
Majestic Company, 58
Martial arts, 136
Marvin Windows & Doors, 18
Master Lock Company, 26
Medeco Security Locks, 32–34, 45
Microwave detectors, 66
Monitors, 96
Motion sensors, 88
MSI Mace, 13, 140
Multiple bolt locks, 26, 41
Murray, Alexander, 79
Myths about security, 1

NAPCO Security Systems, Inc., 67–68
National Association of Private Security Vaults, 61
National Locksmith Association and Trust, 154
National Safeman's Organization, 154
National School of Locksmithing and Alarms, 79
Norden Lock Co., Inc., 41, 44

Outdoor lighting, 85
  motion-activated, 86

Padlocks, 47
Panels, door, 14
Passive alarm, 125
Passive infrared (PIR) detectors, 66
Peripheral devices, 97
Personal safety, 135–141
Police locks, 41–43
Portable alarm, 139
Preso-Matic Lock Company, 48–50
Preventing rape, 136–137
Private vault, 61
Programmable controller, 74
Progress Lighting, 84
Pushbutton locks, 47

Quads, 67
Quick-Vent, 14
Quorum International, Inc., 129, 139
Quorum PAAL, 138

Redi-Exit, 113–115
Reinforcers:
  door, 9
  hinge, 8
Rekeying, 46–47
Relocking devices, 57
Rezek, Julia, 92
Riots, 139–141
Rudolph-Desco, Inc., 12

Safe and Vault Technicians' Association (SAVTA), 154
Safe deposit box, 61
Safes:
  alternatives, 59–60
  buying, 57–59
  fire-resistant, 54
  floor, 55
  in-floor, 56–58
  media, 59
  rating, 54
  specialty, 57
  wall, 55
Safety and security checklist, 161–163
Safety, electric, 81
Sawing, 28
Security consultants, 156
Security film, 22–23
Security services, 152–157
Self-contained alarm, 69
Self-defense, 136–138
Self-defense instructors, 155
Sensors, 64
Sentry Group, 59–60
Sheetglass, 19
Sidelights, 17

Signal processing, 67
"Skeleton key," 26
Smart House integrated system, 74
Smoke detectors, 111
Sourcelist, 197–211
Specialty safes, 57
Steering-wheel locks, 124
Stolen-vehicle retrieval systems, 129
Stop molding, 8
Surveying, 160–161
Surveying an apartment, 161–164
Surviving, 118

Tempered glass, 19
Threat of crime, 4
Thumbscrew devices, 17
Thundarbar, 125
Timers, 86
Touch screen, 77
Traveling safety, 137, 138
Twenty First Century International Fire Equipment & Services Corporation, 116–117

Ultrasonic detectors, 66
Underwriters Laboratories (UL), 54
U.S. Fire Administration, 113
United States Department of Justice, 4

Ventilating wood lock, 19
Vertical deadbolt, 30
Video intercoms, 98–104
  installing, 104
Video Sentry Pan Tilt, 99–103
Viewers, door, 10–12
VSI Donner, 14, 17, 19, 27

Wall safes, 55
Wide-angle viewer, 10–12
Window stickers, 70
Windows, 17–24
  bars, 22
  casement, 18
  double-hung, 18
  glazing, 19–20
  protecting, 22
  replacing, 21–22
  securing, 17–19, 20
  sliding, 17
Wireless alarm, 69, 71
Wrenching, 28

X-10 (USA) Inc., 65, 75–76
X-10 compatible modules, 72
X-10 modules, 165–196

About the Author

Bill Phillips has worked throughout the United States as a locksmith, safe technician, and alarm-systems installer. He is now a security consultant and freelance writer. He is the author or coauthor of four other books, including *Hassle-Free Home Security* (Doubleday Book and Music Clubs) and *Professional Locksmithing Techniques* (Tab Books).

Bill Phillips is a contributing editor of *Home Mechanix* magazine, and his articles appear in *Consumer's Digest, Crime Beat, Locksmith Ledger International, Los Angeles Times, Safe and Vault Technology, Special Report, Your Money,* and many other consumer and professional publications. He is president of the International Association of Home Safety and Security Professionals, and is a certified Residential Protection Specialist.